ADVANCED COAL COMBUSTION

Advanced Energy Conversion Systems
Norio Aria and Stuart W. Churchill (Editors)

Advanced Coal Combustion
Takatoshi Miura (Editor)
ISBN 1-56072-965-1

CONTENTS

EDITORIAL INTRODUCTION

NOVA SERIES "ADVANCED ENERGY CONVERSION SYSTEMS"

Improvement of the efficiency of energy conversion systems and the use of energy in an environmentally friendly way rank among the most important current priorities of mankind.

Significant efforts are being directed in many countries toward the development of advanced energy conversion technologies such as gas turbines, engines and materials, to the determination and the mitigation of the environmental impact thereof, and to the more efficient use of the available energy.

The overall objective of this series of books is not only to provide a fundamental overview of modern energy conversion technologies but also a status report on the latest scientific and technological achievements.

These books will be of interest to students, engineers, teachers and specialists in the field of energy and environmental engineering, aeronautical, chemical and mechanical engineering, and to such related fields as combustion, power stations, furnaces and process heating, and to those operating industrial and propulsive power plants using boilers, gas turbines and steam turbines and automotive engines.

Each volume of this series of books will be edited by world-renowned scientists and engineers. We, as series editors, are deeply indebted to the editors of this series of books for their generous efforts to assure the completion of books of high quality.

Series Editors

Norio Arai, PhD
Professor
Nagoya University, Japan

Stuart W. Churchill, PhD
The Carl V.S.Patterson Professor Emeritus
University of Pennsylvania, U.S.A.

PREFACE

Now 90% of the energy used for transportation, power production, and heating is produced by combustion of liquid, solid and gaseous fuels. To solve a problem of resources depletion, alternative energies (Atomic Energy, Wind Energy, Solar Energy, Biomass Energy, etc.) are studied in the world. However, it is unlikely that this percentage will change significantly for many years. Coal use is expected to increase during the next two or three decades, because coals are still abundant, a low price, and easily available in respect to safety and stability. Thus the study of coal combustion is of continuing importance, especially if we are to conserve our sources of energy and reduce air pollution in a world of increasing population and increasing energy needs. Coal combustion and gasification mechanisms and its application will be very important to use as a main fuel and thus the book is also very important that the present techniques of coal utilization are described by Japanese superior researchers here.

The researchers and engineers intending to study combustion will find that current sources of information fall into two categories, literature on the scientific aspects of combustion and literature on the design and performance of specific technologies such as engines, turbines, and furnaces. Although a considerable number of books on combustion are available, they tend to emphasize the scientific aspects with little detailed information on specific engineering applications. On the other hand, books on applications typically focus on one technology and contain only a few chapters on combustion as applied to that technology.

This book is focused on the most advanced coal combustion. There consists of 9 chapters organized into 3 parts (Coal property, Coal combustion and Application) and is well-balanced constitution contained from fundamentals to applications.

In Chapter 1, a new method is presented to estimate the strength distribution of hydrogen bonds in coal, and changes in hydrogen bond distributions (HBD) with increasing temperature are successfully estimated for all the samples examined. In Chapter 2, the desulfurization behavior in pyrolysis is investigated for raw coals and their density separated samples, more than 90% of organic sulfur has been removed from a subbituminious coal by the combined process of rapid pyrolysis with alkali leaching. In Chapter 3, this article reviews the past research closely relating to prediction of dynamic thermoplasticity, which coals exhibit upon heating.

In Chapter 4, the initiation, behaviors and characteristics of ash deposition under high-temperature reaction conditions are mentioned, and emissions of trace metal compounds in combustion processes and its control technologies are introduced briefly. In Chapter 5, there describes the advanced methods to analyze the distribution of inorganic constituents in coal, the modeling of ash formation and deposition behavior during combustion and the computer tools for predicting slagging and fouling properties. In Chapter 6, there shows the fundamental idea of building a mathematical model for simulation of pulverized coal combustion from a chemical engineering perspective.

In Chapter 7, it has studied and developed a pulverized coal fired power plant for CO_2 recovery as a means to reduce emissions from coal-firing. In Chapter 8, advanced combustion technologies for PCF plants and recent R&D for these technologies are discussed through development of low-NOx combustion technologies. In Chapter 9, there introduce atmospheric fluidized bed combustor and pressurized fluidized bed combustor of coal in Japan.

This book is intended to be a reference volume for researchers, engineers, educators, and graduate students concerned with coal combustion in relation to mechanical and chemical engineering. We hope that they will use this book as a source of enrichment and textual material over the world.

Finally, I wish to express my sincere thanks to all authors and co-authors for their eager efforts. I also express his deep gratitude to Prof. Norio Arai for the interest shown in this manuscript and to Dr. Tsuyoshi Yamamoto for the work done by him in editing the book.

Takatoshi Miura

I. Coal Properties

Chapter 1

IN SITU FTIR MEASUREMENT OF HYDROGEN BOND DISTRIBUTIONS IN COAL DURING HEAT TREATMENT

Kouichi Miura[*] *and Kazuhiro Mae*[*]

1. INTRODUCTION

There are believed to exist five associative interactions in coals [1]. These are covalent bonds, ionic linkages, hydrogen bonds, van der Waals forces, and π-π interactions. Of the five associative interactions hydrogen bonds, noncovalent associative interactions, are believed to play a key role to keep the macromolecular structure of low rank coals [1, 2, 3, 4, 5, 6, 7], and hence utilization of the low rank coals is significantly affected by the hydrogen bonding. Since a large amount of water is retained in the lower rank coal due to the hydrogen bonds, drying is one of the drawbacks for the utilization of the coal. Hydrogen bonded carboxylic and hydroxylic groups in the coal cross-links at the early stage of pyrolysis significantly affect the subsequent main pyrolysis reactions [8]. Usually the crosslinking reactions suppress the formation of volatiles. In solvent extractions of coal and/or the liquefaction of coal the interactions of coal surface and solvents play a crucial role. Fundamental properties such as physical properties, i.e., glass transition temperature, are significantly affected by the existence of hydrogen bonds. Therefore it is very important to quantify the hydrogen bonds in coal from both practical and fundamental viewpoints. Quantification means the estimation of the amount and the strength of each hydrogen bond.

Several attempts have been made to estimate the hydrogen bonds in coal using various methods, including solvent swelling phenomenon of coal, DSC, NMR, and FTIR. Experiments and analyses of solvent swelling of coal have been performed in relation to the role of noncovalnet bonds in coal. One of the pioneer works was the one performed by Sanada and Honda [9]. They estimated the density of cross-links and the mean molecular weight between two adjacent cross-links by applying the Flory-Rehler theory

[*] Department of Chemical Engineering, Kyoto University, Kyoto 606-8501, Japan (miura@cheme.kyoto-u.ac.jp).

for polymers to the swelling of coal by pyridine vapor. Since this pioneer work many works have been presented, which have contributed to clarify the coal structure. Recently Larsen et al., for example, have presented a method to estimate the number of hydrogen-bond cross-links in coal by utilizing the swelling of coal with a polar solvent diluted by nonpolar solvent [10]. This method clarified that there are about 1.7 hydrogen-bond cross-links per 100 carbon in Illinois No. 6 coal and 0.3 hydrogen-bond cross-links per 100 carbon in Pittsburgh No. 8 coal. However, no information on the amount and strength of hydrogen bonds in coal has been determined from solvent swelling examinations.

Direct or indirect estimation of heat accompanying the formation or disruption of hydrogen bonds is another method to estimate the strength of hydrogen bonds. Lucht et al. [11], Mackinnon et al. [12], and Yun et al. [13], for example, examined the glass transition of coal by the DSC measurement. Mackinnon et al. observed the 2nd order phase glass transition at around 110 °C during the heating of various coals. However, direct calorimetric measurements have been in general not successful, because many factors have to be taken into account to convert the experimentally obtained heat into the strength of hydrogen bonds.

Heats for wetting of coal in solvents such as water, methanol, etc. have been measured calorimetrically to elucidate oxygen functional groups in coal. As early as 1957 Iyenger and Lahiri and Mazumdar et al. tried to correlate the heat of wetting and the energy of hydrogen bonding between coal oxygen functional groups and methanol [14]. To calculate the energy of hydrogen bonds, they assumed that each COOH group is capable of hydrogen bonding with two CH_3OH molecules and that OH and C=O groups are each capable of hydrogen bonding with one CH_3OH molecule, respectively, and assigned the following hydrogen bond energy value for each hydrogen bond formed by these functional groups: COOH: 8.2 kcal/mol, C=O: 7.0 kcal/mol, and OH: 6.2 kcal/mol [15]. By further assuming that all C=O groups are either chelated or interlocked with their corresponding number of OH groups and therefore not available for hydrogen bonding with water, which was deduced from the strong IR absorption band near 1600 cm^{-1} in all coals, they found a remarkable statistical agreement between the heat of wetting and the total energy of hydrogen bonding calculated from the amount of oxygen functional groups and the hydrogen bond energies given above as shown in **Figure 1**. Similar correlations were obtained for the heats of wetting in water and chlorobenzene and corresponding hydrogen bond energies, where the hydrogen bond energies for OH···O and OH···Cl were respectively assumed to be 4.5 kcal/mol and 3.9 kcal/mol. These pioneer works are very interesting in the sense that they related the hydrogen bonding energy and a directly measurable quantity, heat of wetting, although it may be questionable if their interpretation is correct.

Figure 1 Relationship between heat of wetting and modified energy of hydrogen bonds [14].

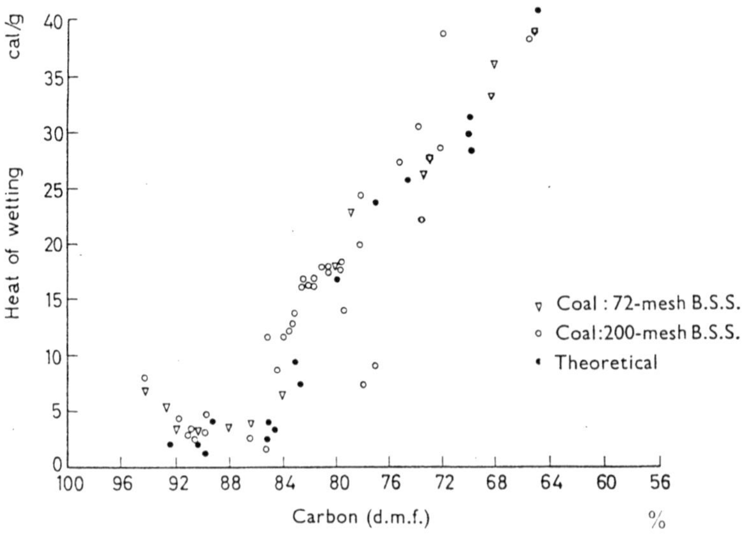

One of the indirect methods for the estimation of heat associated with hydrogen bond is the estimation of isosteric adsorption enthalpy (heat of adsorption) from the adsorption isobars. Allardice and Evans estimated the heat of adsorption of water during the course of desorption of water from a raw brown coal [16]. It increased from the level of heat of vaporization (43.88 kJ/mol-H_2O at 25 °C) to 60 kJ/mol-H_2O with the decrease of the amount of water adsorbed. This showed that coal has different adsorption sites for water; in other words, coal oxygen functional groups have different hydrogen bonding abilities. When the heat of adsorption was estimated from the readsorption isobars for the dried coal, on the other hand, its value was almost same as the heat of vaporization irrespective of the amount of water adsorbed. This result shows that most of hydrogen bonds formed among the oxygen functional groups in the dried coal are stronger than the hydrogen bonds that can be formed between the oxygen functional groups and water. This shows that the strength of hydrogen bonds in the dried coal cannot be estimated from water adsorption data. To overcome this drawback, Glass and Larsen presented a method to separate the heats of adsorption of organic bases on coal into nonspecific and specific interactions [17]. The specific interaction heats, which are judged to be the interactive energies between the bases and coal oxygen functional groups, are found to be equal to the heats of hydrogen-bond formation between those bases and p-fluorophenol in liquid phase [18] as shown in **Figure 2**. This clarified that the base-coal interaction and base-phenol interaction in liquid phase can be treated in the same way. This analysis method would present the possibility to estimate the strength distribution of hydrogen bonds in coal by changing the bases systematically.

Figure 2 Specific adsorption enthalpy, $q_{st,specific}$, on demineralized Illinois No.6 coal vs. heat of hydrogen-bond formation with p-fluorophenol, $\Delta H_{f,PFP}$, for seven bases [17].

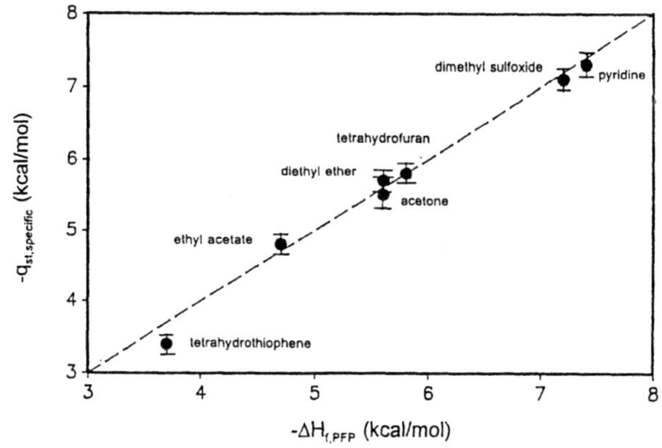

Infrared (IR) spectroscopy is surely one of most promising methods to quantify the hydrogen bonds in coal directly [19, 20, 21, 22, 23, 24, 25, 26, 27, 28, 29]. As Painter et al. reviewed [21], Fourier transform infrared spectroscopy with an on-line microcomputer has opened a new era of IR spectrometry. It has become routine to measure IR spectra and to manipulate them by spectral subtraction, curve resolving, factor analysis, etc., which lead IR spectroscopy to become a quantitative analysis method. However, there are several problems to be solved to apply the FTIR method to quantitative analysis of coal. The most commonly employed sample preparation method is the so-called KBr pellet method. A small amount of coal (1 to 3 wt%) is mixed with KBr powders, and they are pelletized to a transparent disk that is served to transmittance IR measurement. It is essential to grind both coal particles and KBr particles into the size less than the wavelength of the IR radiation. The most severe difficulty is the ubiquitous presence of water. It seems to be almost impossible to remove the water effect especially when we intend to estimate the amount of OH groups from the OH stretching absorption bands around 3000 to 3600 cm^{-1}, since strong and broad water adsorption bands appear in the region. It is also very difficult to do subtractions of spectra on a 1/1 basis to obtain a difference spectrum without proper internal standards, because it is impossible to prepare two disks under perfectly identical conditions.

Overcoming these problems, Solomon et al. tried to quantitatively estimate the amounts of OH [23], aliphatic hydrogen and aromatic hydrogen [30] using the KBr pellet method in a series of works where careful sample preparation and elaborate data analysis were performed. It is reported that drying the KBr pellet for 48 h at 105 °C is enough to eliminate KBr-H_2O bands and that the absorptivity at 3200 cm^{-1} correlates very well with the total amount of OH irrespective of coal rank. On the other hand, the absorptivities for aliphatic hydrogen, a_{al}, and aromatic hydrogen, a_{ar}, were found to be coal rank dependent. Their values estimated respectively from the sets of peaks near 2900 cm^{-1} and near 800 cm^{-1} were a_{al} = 746 and a_{ar} = 686 for bituminous coals and a_{al} = 710 and a_{ar} = 541 for subbituminous coals and lignites.

Solomon et al. tried to estimate the amount of OH directly from the OH stretch region to simplify the estimation, Painter et al. proposed to measure the spectrum ranging from 1550 to 1900 cm^{-1} of the acetylated coal for the estimation of OH amount [24]. By this method acetylated phenolic OH, alkyl OH, and NH groups were estimated from the bands of 1770, 1740, and 1670 cm^{-1}, respectively. This method was employed to overcome a number of problems associated with the use of the absorption band near 3400 cm^{-1} as a direct measure of OH group content. The main problems are water adsorption by the KBr matrix, water bound to coal in a complex manner, and the presence of more than one band in this region of the spectrum. These factors were judged to conspire to make quantitative measurements based on the 3400 cm^{-1} band suspect.

There are several problems to quantify the absorption bands near 3400 cm^{-1} as stated above, but the absorption bands ranging from 2400 to 3600 cm^{-1} contain important information on OH group associated hydrogen bonds. Fuller and Smyrl [25] showed that the *in-situ* diffuse reflectance IR Fourier transform (DRIFT) technique with neat, undiluted, coal samples can be well utilized to trace the *in-situ* reactions such as oxidation, dehydration, etc. [22] This technique seemed to resolve most of the problems associated with the KBr pellet method: no sloping base line appears; water adsorption by KBr does not occur; it is very easy to do subtractions of spectra on 1/1 basis to obtain a difference spectrum. Analyzing difference spectra obtained by this technique, Fuller and Smyrl clarified the attributions of the absorption bands ranging from 2800 to 3625 cm^{-1} as follows:

3625 cm^{-1};	weakly hydrogen bonded entities
3540, 3390, 3290 cm^{-1};	structural difference and various types of hydrogen bonds
2850 - 3000 cm^{-1};	aliphatic hydrogen
3000 - 3100 cm^{-1};	aromatic hydrogen

Larsen and Baskar related the strength of coal-solvent interactions with the most direct spectroscopic probe available: changes in the hydroxyl stretching frequency on hydrogen bond formation [31]. The spectrum of vacuum dried coal mixed with KBr was measured using a controlled-atmosphere cell by the DRIFT technique. The coal was then exposed to deuteriated solvent vapor, and the spectrum of the coal adsorbing the solvent was measured. The value of OH stretching frequency shift was estimated from the difference spectrum of the two spectra. **Figure 3** shows the wavenumber shift against the pK_a value of the solvents (bases) used. The linear relationship was similar to that obtained when similar bases form hydrogen bonds to phenol in liquid phase. The good correlation in Fig. 3 shows that this technique holds the promise of providing quantitative information of relative hydrogen-bond strengths.

Painter *et al.* [26] reviewed a wealth of literature concerning hydrogen bonds, and attempted to define explicitly the types of hydrogen-structure present. They clarified that the functional groups present in coals and are capable of forming hydrogen bonds with various acceptor groups are hydroxyls, and phenolic and carboxylic acids, and assigned the OH stretching frequency bands to the following hydrogen bonds:

3611 cm^{-1};	free OH groups
3516 cm^{-1};	OH-π hydrogen bonds
3400 cm^{-1};	self associated n-mers (n>3)
3300 cm^{-1};	OH-ether O hydrogen bonds
3200 cm^{-1};	tightly bound cyclic OH tetramers
2800 - 3100 cm^{-1};	OH-N (acid/base structures)

Figure 3 Plot of frequency shift of the O-H stretching band of Rawhide coal vs. base strength of the adsorbed solvent [31].

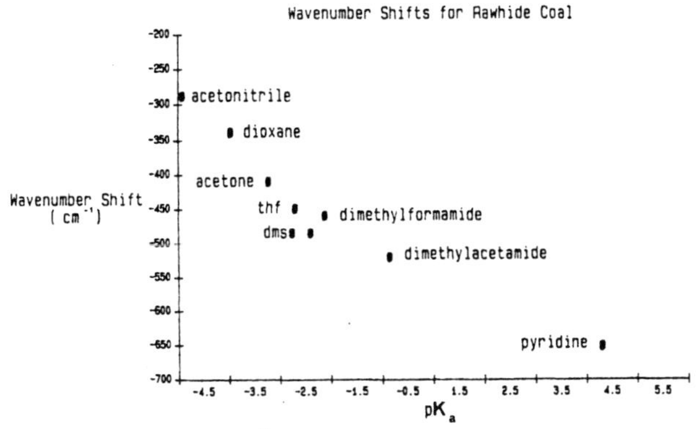

Figure 4 shows a typical spectrum of a coal extract that was used to identify the bands. Existences of the bands were determined by a curve resolving method based on the second derivative of the spectrum and the eye and brain examination. The OH-π hydrogen bonds are judged partly to be due to OH groups that sterically can not find an energetically preferable partner. This implicitly meant that very few OH groups exist as free OH. They also clarified that the frequency shifts from that of the free OH and the intensities of the bands provide fundamental parameters that can be used to determine structural characteristics such as bond lengths and the enthalpy of bond formation, although they did not estimate them.

As briefly reviewed, it was clarified that there are several types of hydrogen bonds in coal, and that their existences are most vividly shown by the OH stretch frequency shift of IR absorption bands. However, no definite methods have not been presented to estimate the strength distribution of hydrogen bonds probably because of the difficulty in analyzing the OH stretch frequency bands associated with water bound to coal in a complex manner and the presence of more than one band in the region of the spectrum. If the KBr pellet method was employed, water adsorption by the KBr matrix should also be taken into account.

Figure 4 FTIR spectrum of a cast film of a pyridine extract of a Polish coal used for identifying different hydrogen bonds [26].

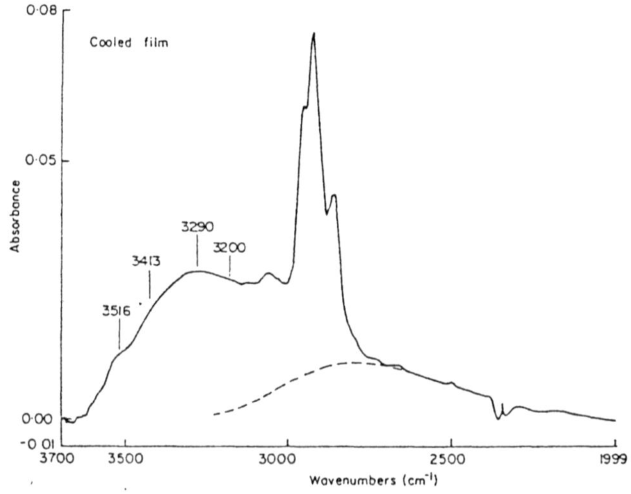

Apart from coal hydrogen bonds there have been a wealth of works on various hydrogen bonds, including various reviews and several books [32, 33, 34]. In liquid phase hydrogen-bonded adduct formation reactions between the OH functional groups of phenols and/or alcohols and various bases have been examined in detail using FTIR and thermodynamic considerations. The reaction for a phenol, for example, is represented by the following stoichiometric equation:

$$\text{Ph-OH} + \text{B} \rightarrow \text{Ph-OH}\cdots\text{B} \; ; \; \Delta H \qquad (1)$$

where Ph-OH···B represents the hydrogen-bonded adduct, and ΔH is the enthalpy change of the reaction. The ΔH values for various combinations of phenols and bases were estimated by applying the van't Hoff equation to the equilibrium constants measured. The amount of hydrogen bonds is also obtained in the course of the measurement of equilibrium concentrations. When the hydrogen bond is formed, the IR wavenumber of the O-H stretching vibration of the phenol shifts to a low wavenumber and generates heat ($\Delta H < 0$). The relationships between the enthalpy change, ΔH, and the OH wavenumber shif t, $\Delta \nu_{OH}$, were examined for various phenol-base and alcohol-base combinations [18, 35, 36, 37, 38, 39], and a linear relation was found to hold between ΔH and $\Delta \nu_{OH}$ by many investigators [38, 40]. The relationship is well known as "linear enthalpy-spectroscopic shift relations". The relationship obtained by Drago et al. [38], for example, from many measurements (**Figure 5**) is given by

$$-\Delta H = 0.067\,\Delta \nu_{OH} + 2.64 \;\; [\text{kJ/mol}] \qquad (2)$$

Figure 5 A plot of $\Delta \nu_{OH}$ (cm^{-1}) vs. - ΔH (kcal/mol) for different phenol-base combinations: ○, *p-t*-butylphenol as acid; □, phenol as acid; Δ, *p*-chlorophenol as acid; ⬡, *m*-trifluoromethylphenol as acid [38].

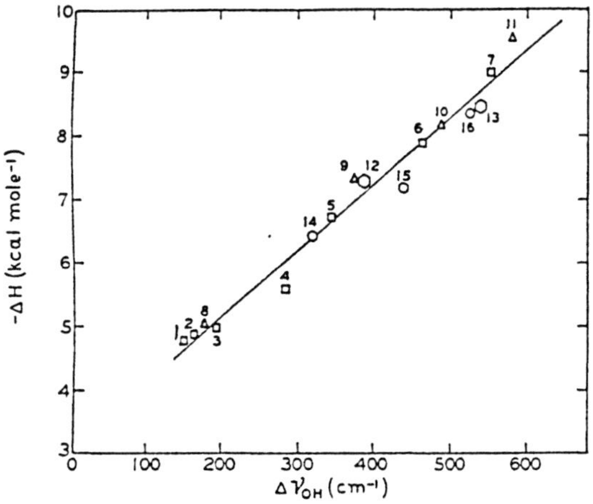

The values of - ΔH was interpreted as the strength of hydrogen bond by many investigators [10, 11, 12, 13]. However, Drago et al. thought that ΔH consists of two contributions: the decrease in the O-H bond energy due to the formation of the hydrogen-bonded adduct, δD_{OH} (> 0), and the bond dissociation energy of hydrogen bond formed, D_{HB}, as

$$\Delta H = \delta D_{OH} - D_{HB} \tag{3}$$

They thought that D_{HB} is an index better than ΔH for the strength of the hydrogen bond. Then a relationship between δD_{OH} and $\Delta \nu_{OH}$ is required to obtain the relationship between D_{OH} and $\Delta \nu_{OH}$. They approximated the stretching of the hydrogen-bonded OH by the anharmonicity stretching of free OH, and related δD_{OH} and $\Delta \nu_{OH}$ theoretically as follows:

$$\delta D_{OH} = (hcN / 4\chi_e)\, \Delta \nu_{OH} = D_{OHf}\,(\Delta \nu_{OH} / \nu_e) = 0.131\, \Delta \nu_{OH} \tag{4}$$

where h is Plank's constant, c is the speed of light, D_{OHf} is the bond dissociation energy of free O-H (= 420kJ / mol) and ν_e is the wavenumber of harmonic vibration of the free OH in cm^{-1}, and χ_e is the anharmonicity constant for the oscillator.

Using eqs. (3) and (4), eq. (2) can be solved to obtain D_{HB} as follows:

$$D_{HB} = 0.198\, \Delta \nu_{OH} + 2.64 \ \text{[kJ/mol]} \tag{5}$$

Now either eq. (2) or (5) directly relates the OH wavenumber shift to the strength of hydrogen bonds for various phenol-base and alcohol-base hydrogen bonds in liquid phase.

2. New Method Developed for Estimating Strength Distribution of Hydrogen Bonds in Coal [41, 42]

2.1. Definitions and Basic Concepts of the New Method

We have attempted to develop a method to quantify the hydrogen bonds in coal based on the findings reviewed above. The hydrogen bond formation reaction of OH groups can be defined by following eq. (1) as follows:

$$\text{Coal-OH} + \text{B} \rightarrow \text{Coal-OH}\cdots\text{B}\ ;\ \Delta H \tag{6}$$

where Coal-OH represents a hypothetical state in which the OH exists as a free OH group. The bases, B, in this case represent O and N in OH, COOH, C=O, and N-H groups.

We decided to analyze the OH stretching absorption bands, because the information on the hydrogen bonds is concentrated on the absorption bands. To do so, several problems must be solved: how do we eliminate the effect of adsorbed water interacting with coal intimately?; how do we determine the amount of each hydrogen bond from the complex adsorption band?; are eqs. (2) and (5) applicable to the hydrogen bond in coal? The water problem associated with KBr was overcome by using the DRIFT technique with neat coal samples.

The effect of adsorbed water was not intended to be eliminated but intended to be included in the analysis, since the adsorbed water significantly affects the hydrogen bonds in coal. Then the following hydrogen bond formation reaction was taken into account in addition to the reaction (6).

$$\text{HO-H} + \text{B'} \rightarrow \text{HO-H}\cdots\text{B'}\ ;\ \Delta H' \tag{7}$$

where HO-H again represents a hypothetical state in which the OH exists as a free OH group. The bases, B', in this reaction involves O in water in addition to the bases in coal. This treatment apparently looks to make the analysis complicated, but in reality this facilitated the analysis as stated below.

The amounts of individual hydrogen bonds, in other words the amounts of OH contributing to different hydrogen bonds, were determined by the curve resolving method. First, the absorption band ranging from 2400 to 3750 cm^{-1} was regarded to consist of the following 10 absorption bands, 7 OH stretch bands and 3 CH stretch bands.

3640 cm^{-1};	free OH groups
3530 cm^{-1};	OH-π hydrogen bonds
3400 cm^{-1};	self associated n-mers (n>3)
3280 cm^{-1};	OH-ether O hydrogen bonds
3150 cm^{-1};	tightly bound cyclic OH tetramers
2940 cm^{-1};	OH-N (acid/base structures)

2640 cm^{-1};	COOH dimers
3050 cm^{-1};	aromatic hydrogens
2993, 2920 cm^{-1};	aliphatic hydrogens

The peak positions are a little different from those proposed by Painter et al. [26], and the hydrogen bonds attributable to COOH dimers were also taken into account judging from the second derivative of the spectrum and the brain and eye examination. **Figure 6** schematically shows the hydrogen bonds taken into account in this work. Assuming the Gaussian distribution for each absorption band, the spectrum was curve resolved to obtain the intensity of each band.

Figure 6 Schematic representation of hydrogen bonds assumed to exist in coal in this work.

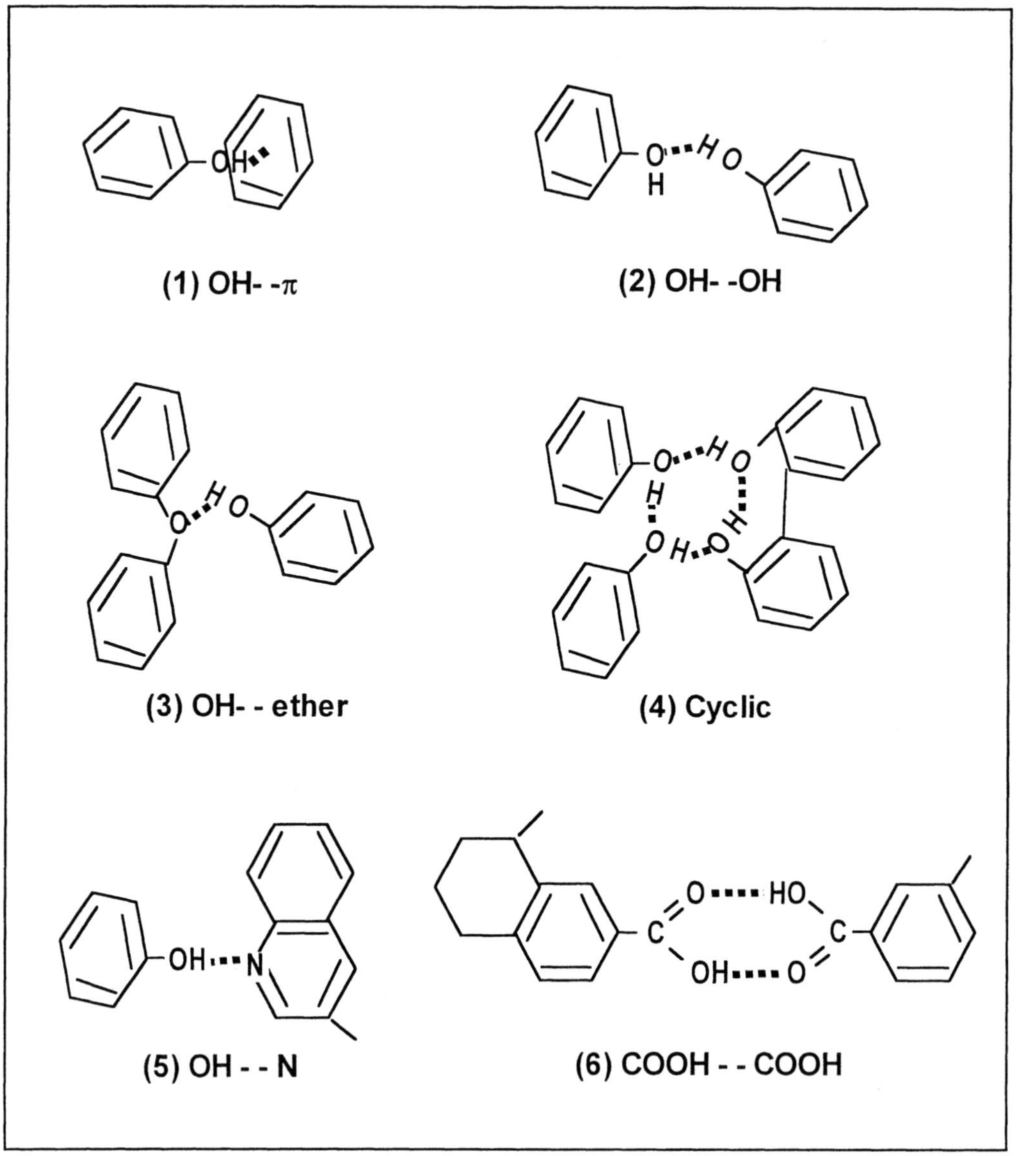

Next, the absorptivity (extinction coefficient) of each band must be determined to convert the individual intensities to corresponding amounts of OH groups. It is known that the absorption intensity of OH stretch frequency forming a hydrogen bond increases with the increase of frequency shift, $\Delta\nu_{OH}$. Therefore, it is necessary to estimate the absorptivity of each absorption band independently in principle. Detoni et al. examined the change of absorptivity of the NH stretching band of isothiocyanic acid (HNCS) accompanying the formation of hydrogen bonds with various acceptors in CCl_4 systematically [43]. They found the following relationship between the absorptivity of the hydrogen bonded NH, α_{NH}, and that of free NH, $\alpha_{NH,0}$:

$$\alpha_{NH} = \alpha_{NH,0}(1+0.0141\Delta\nu_{NH}) \tag{8}$$

where $\Delta\nu_{NH}$ is the NH wavenumber shift. This relationship shows that the absorption intensity of the hydrogen bonded NH increases in proportion to $\Delta\nu_{NH}$. Based on this work we arbitrarily assumed a similar relationship for the change in absorption intensity of hydrogen bonded OH, and estimated the proportional constant from the intensity and the $\Delta\nu_{OH}$ values measured for several model compounds (Benzoic acid and 2-Naphtol) as

$$\alpha_{OH} = \alpha_{OH,0}\,(1+0.0147\Delta\nu_{OH}) \tag{9}$$

where α_{OH} is the absorptivity of the hydrogen bonded OH and $\alpha_{OH,0}$ is the absorptivity of the free OH. The proportionality constant obtained for the OH associated hydrogen bonds (eq. (9)) was found to be very close to that for the NH associated hydrogen bonds (eq. (8)). Thus we can calculate the amounts of OH contributing to different hydrogen bonds using eq. (9) when we can estimate $\alpha_{OH,0}$.

The last problem must be solved is if eqs. (2) and (5) are applicable to the hydrogen bond in coal. Their applicability is supported by the works of Larsen et al. (Figs. 2 and 3): Figure 2 qualitatively shows that the OH shift increases with base strength as in the case for phenol-base hydrogen bonds, and Figure 3 suggests the equivalence of the ΔH (= $q_{st,specific}$) values and $\Delta H_{f,\,PFP}$ values. These results suggest that the coal hydrogen bonds are treated in a similar way as the phenol-base hydrogen bonds. Although not directly related the coal hydrogen bonds, Kiselev [44] found that a linear relationship between the heat of adsorption, ΔQ_a, and the $\Delta\nu_{OH}$ for a hydroxylated silica surface and various vapors including steam. It was represented by

$$\Delta Q_a = 0.064\,\Delta\nu_{OH} - 1.16 \ \ [\text{kJ/mol}] \tag{10}$$

These works clarified that the OH wave number shift caused by the formation of hydrogen bonding is well related to the enthalpy change of the hydrogen bonding formation. Based on these considerations we have judged that eqs. (2) and (5) are applicable to the hydrogen bond in coal as a first approximation. Using eqs. (2) and (5), we can now estimate the strength of individual hydrogen bonds in coal from the $\Delta\nu_{OH}$ values.

2.2. Procedure to Estimate the Strength Distribution of Hydrogen Bonds

The procedure to obtain the strength distribution of hydrogen bonds by the proposed method is summarized as follows:

1. Divide FTIR spectrum ranging from 2200 to 3700 cm^{-1} into 6 hydrogen bonded OH bands, 3 CH stretch bands, and the free OH band, if it existed, by a curve fitting method as schematically given in **Figure 7(a)**. The Gaussian distribution was assumed for each absorption band in this work. Three CH stretch bands are easily eliminated from the spectrum.

Figure 7 Procedure to obtain the strength distribution of hydrogen bonds by the proposed method.

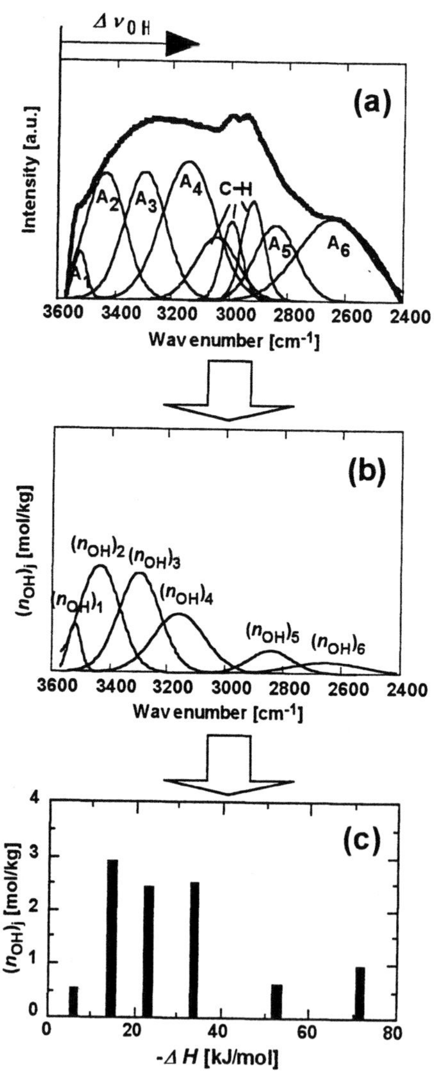

2. Estimate the amount of OH for the j-th peak, $(n_{OH})_j$, by the Beer law as follows:

$$(n_{OH})_j = (1/\alpha_j)A_j \tag{11}$$

where A_j and α_j are respectively the integral intensity and the absorptivity of the j-th peak. In this analysis α_j was assumed to be represented by eq. (9).

$$\alpha_j = \alpha_0\{1+0.0147(\Delta v_{OH})_j\} \tag{12}$$

where α_0 is the absorption coefficient of the stretch vibration of the free OH. Then the amount of the OH corresponding to the j-th peak, $(n_{OH})_j$, is given by

$$(n_{OH})_j = \frac{1}{\alpha_0\{1+0.0147(\Delta v_{OH})_j\}} A_j \tag{13}$$

3. Calculate the strength of the j-th hydrogen bond, $(-\Delta H)_j$ by eq. (2) as

$$-\Delta H = 0.067\,\Delta v_{OH} + 2.64\ \text{[kJ/mol]} \tag{2}$$

   The value of $(D_{HB})_j$ can also be obtained from eq. (5) if necessary.
4. $(n_{OH})_j$ vs. $(-\Delta H)_j$ relationship gives the distribution of the strength of hydrogen bonds as shown schematically in Fig. 7(c).

Table 1. Ultimate analyses and oxygen distribution of 8 coals used [wt%]

Coal (abbre.)	Ultimate analysis [d.a.f.]				Oxygen content [d.a.f.]			
	C	H	N	O(diff.)	as COOH	as OH	as C=O	as Ether
Morwell (MW)	67.1	4.8	0.7	27.4	7.97	5.92	-	-
Beulah-Zap (ND)	72.9	4.8	1.2	20.1	3.81	9.16	1.96	2.28
Wyodak (WY)	75.0	5.4	1.1	21.0	3.33	7.68	0.74	2.56
Illinois #6 (IL)	77.7	5.0	1.4	13.5	0.23	5.86	0.93	0.57
Blind Canyon (UT)	80.7	5.8	1.4	13.5	0.23	5.22	0.63	3.68
Pittsburg #8 (PITT)	83.2	5.3	1.6	8.8	0.16	4.32	0.18	2.08
UpperFreeport (UF)	85.5	4.7	1.6	7.5	0.14	1.96	0.44	0.36
Pocahontas (POC)	91.1	4.4	1.3	2.5	0.05	0.98	0.32	0.24

3. Experimental

3.1. Samples

Seven Argonne premium coals and an Australian brown coal (Morwell) were used as coal samples. The ultimate analyses and the amount of OH and COOH oxygen [45, 46] are given in **Table 1**. Coal samples -200 mesh were further ground for 1 minute in a mortar just before measurement. As model compounds for coal an ion exchange resin (Mitsubishi Chemicals, WK11) that has carboxylic groups as ion exchange sites (**Figure 8**) and a lignin (Nakalai Tesque, Co.) were also used. The amount of COOH in the resin was estimated to be 10.3 mol/kg by [13] C-NMR measurement.

Figure 8 Structure of an ion exchange resin used as a model compound for coal.

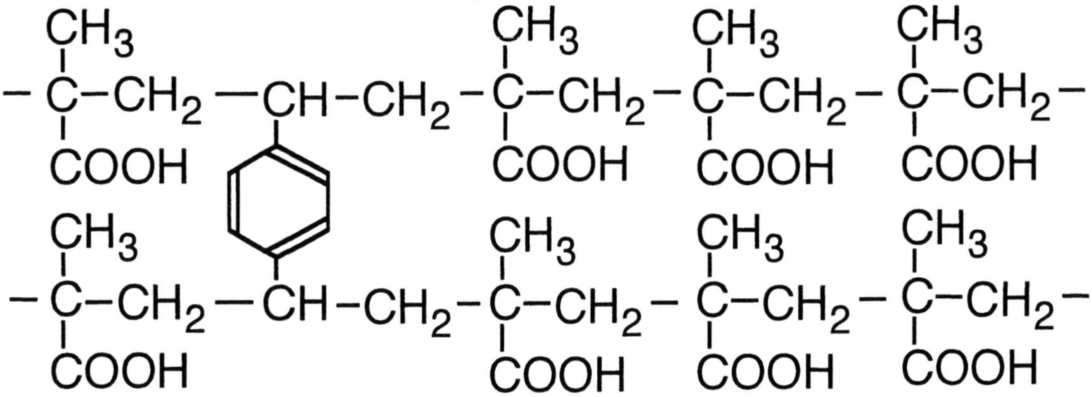

3.2. Apparatus and Procedure

All the IR spectra were measured at 4 cm^{-1} resolutions on a JEOL JIR-WINSPEC50 FTIR spectrometer with a microscope stage. A Mettler 84HPT hot stage, shown schematically in **Figure 9**, was attached to the microscope stage for the *in-situ* measurement. A mirror, which is believed not to absorb water, was used as a background. About 0.5 mg of neat coal samples were leveled and pressed on the mirror by hand using a spatula. It is essential to make a flat surface for obtaining good and reproducible spectra. The coal samples were equilibrated at 30 °C in a flowing nitrogen stream. Then they were heated at the rate of 5 K/min up to 300 °C during which ultra high purity nitrogen was continually supplied at the rate of 100 cc/min to prevent any possible leakage of air into the chamber and to keep the stability of the temperature. A BaF_2 disc was put on the pinhole from where the light passes through. The light source was a heated Ni-Cr wire and the detector was MCT (mercury cadmium-tellurium). FTIR spectrum was collected by acquisition of 100 scans at every 20 °C of interval.

Figure 9 Schematic of a modified Metler 84HPT hot stage used for *in-situ* FTIR measurements.

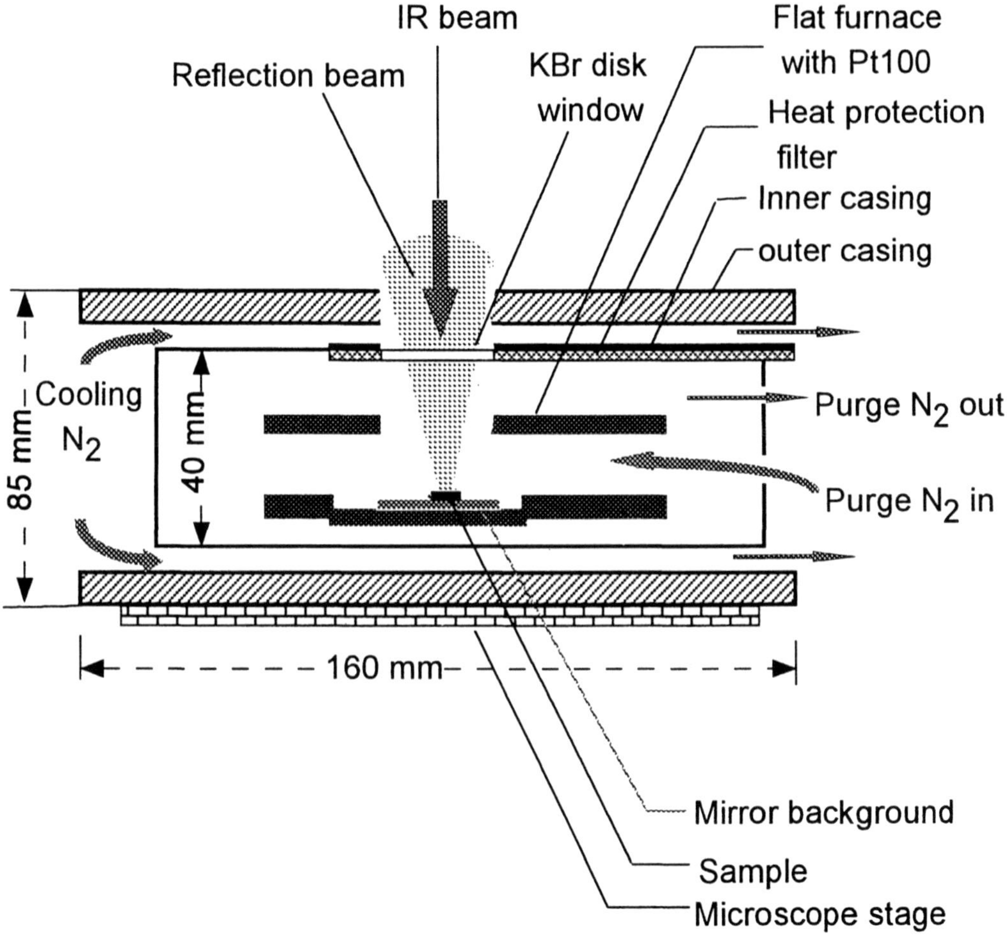

4. Results and Discussion

4.1. Validity of *In-Situ* DRIFT Technique Employed

As discussed above, there are several experimental problems to be solved for analyzing FTIR spectrum quantitatively. However, most of problems could be solved by using the in-situ DRIFT technique. The water adsorption problem was also overcome by analyzing the spectrum measured with the adsorbed water. **Figure 10** shows the change of the FTIR spectra for the resin with the increase of temperature. One of the problems encountered by many investigators in using *in-situ* FFTIR was the spectra quality at high temperatures due to the change in refractive index. Very beautiful spectra with almost flat base lines show that such a problem is well overcome by the DRIFT technique employed here. To examine the change of the OH absorption band region, the spectra ranging from 2400 to 3650 cm^{-1} were enlarged and shown in **Figure 11**. The weight change and the gas formation rates measured under the same experimental conditions using a TG-MS method are also shown in Fig.11 to clarify the change during the heat up of the resin. When the resin was heated from room temperature to 140 °C, the intensity of the absorption bands ranging from only 2800 to 3400 cm^{-1} decreased. This decrease can be well associated with the desorption of adsorbed water referring to the water formation profile in the lower figure. This also indicates that water is adsorbing with OH groups by rather weak hydrogen bonds. By further heating all OH associated absorption bands decrease significantly, and these decreases accompany again the formation of water as shown in the lower figure. The water formed in this temperature range is judged to come from the decomposition reactions of carboxylic groups represented by

$$2\text{-COOH} \rightarrow \text{-COOCO-} + H_2O \qquad (14)$$

This can be confirmed by the appearance of distinct absorption bands of an anhydride (-COOCO-) at near 1800 cm^{-1} as shown in Fig. 10. Thus it was found that the changes of the OH absorption bands measured by the *in-situ* DRIFT technique give a lot of information, including the reaction of the early stage of pyrolysis. This will be shown more distinctively through the quantitative analysis made below.

Figure 10 Change in the *in-situ* FTIR spectrum of the ion exchange resin (WK11) with the increase of temperature.

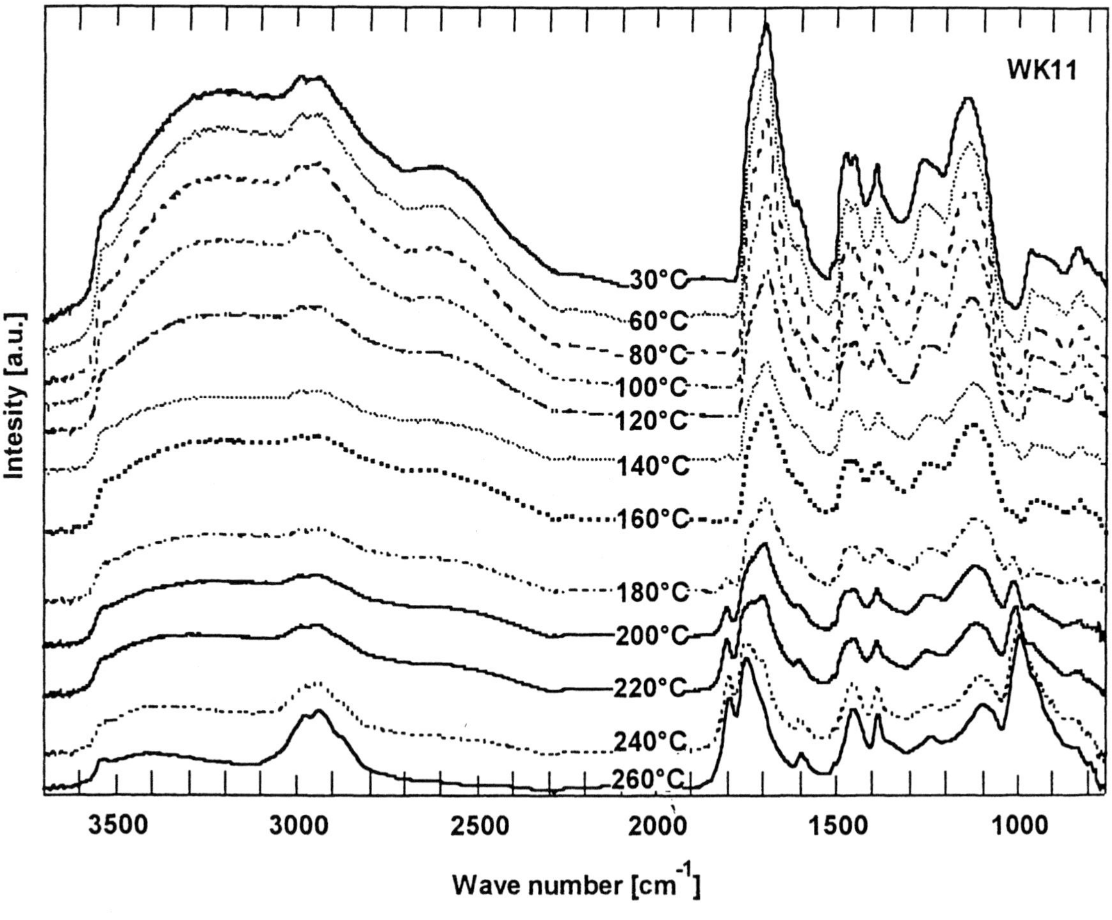

Figure 11 Changes in the OH stretching bands, weight and gas formation rates during a heating of the ion exchange resin (WK11).

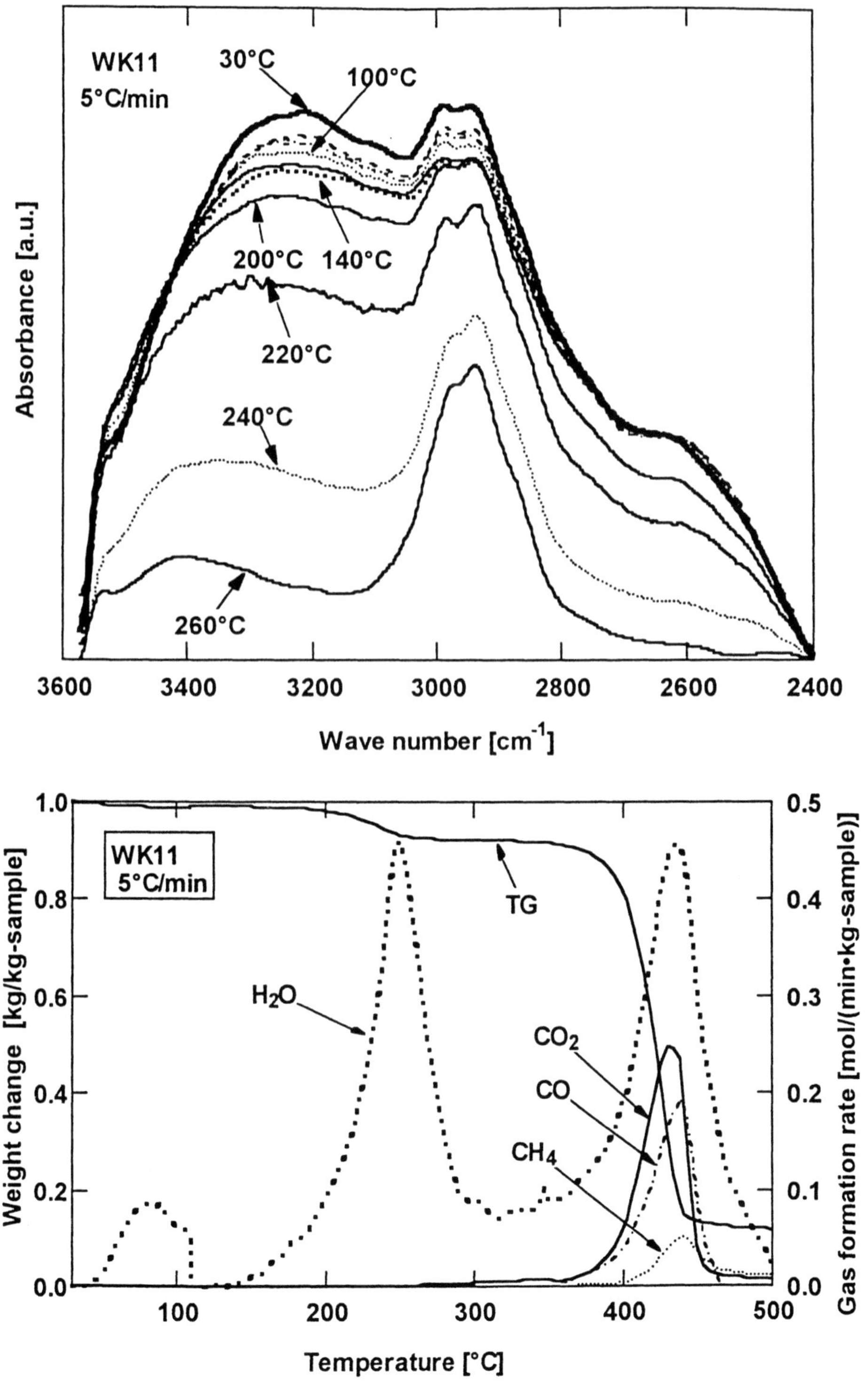

4.2. Change of OH Absorption Bands of Coal with Heating

Figure 12 shows the *in-situ* FTIR spectra ranging from 2200 to 3700 cm^{-1} measured for the eight coals. Very beautiful and reproducible spectra could be obtained by use of the proposed measurement technique for all the coals. Very flat base lines were obtained at all the temperatures, and the spectra for POC were almost the same above 70 °C. This suggested that the spectra were not affected by the temperature below 300 °C. Separate experiments were performed using kaolinite to check the change of spectra with increasing temperature up to 350°C. The mirror backgrounds at 30°C and at corresponding temperatures were used respectively to get the FTIR spectra. The spectra did not change with the background at different temperatures. Then all the spectra shown in this work were obtained using the background at 30°C. This made the *in-situ* measurement convenient, fast and reliable.

Figure 12 *In-situ* FTIR spectra measured at every 20 °C for seven Argonne premium coals and an Australian brown coal.

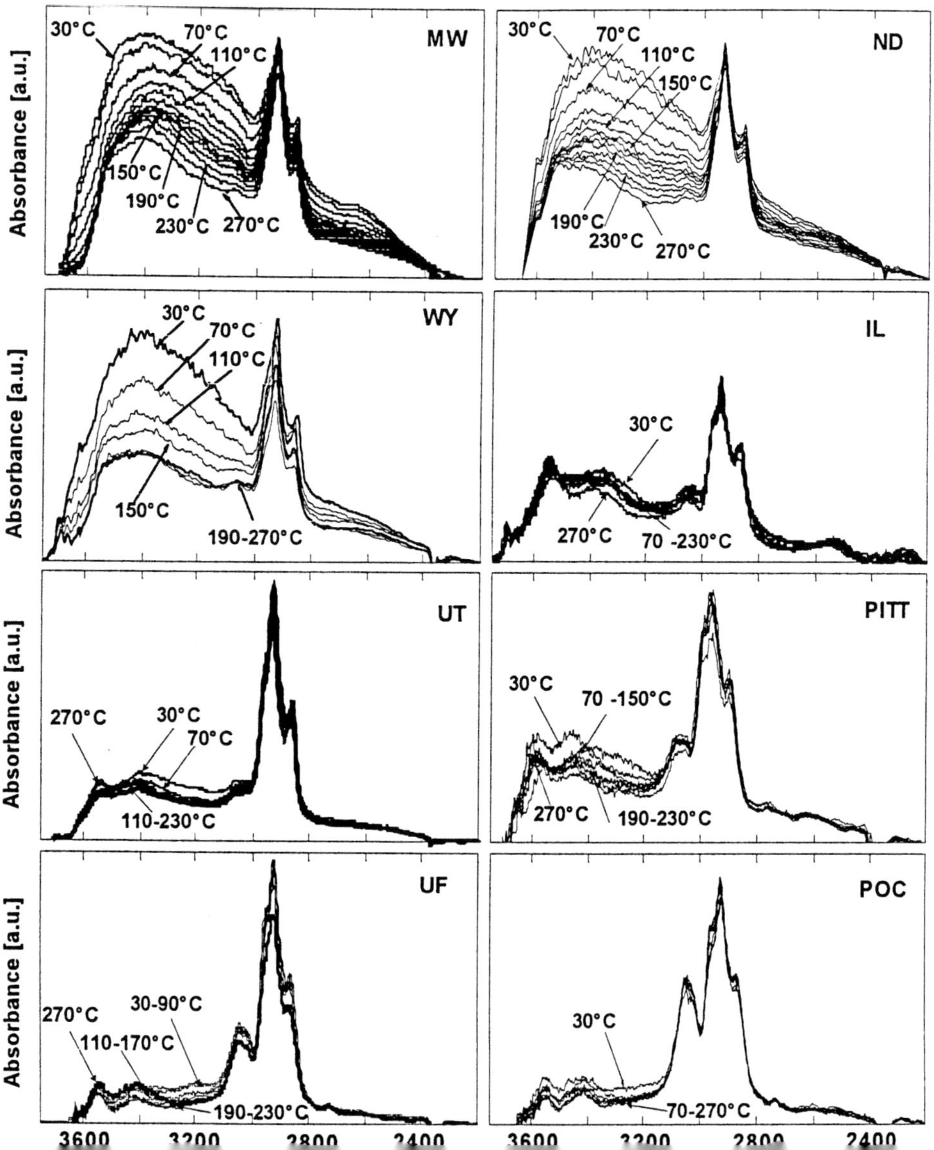

To get accurate information of hydrogen bonded OH groups, it is essential to ensure that the artifact does not exist during experiments and/or during the data processing after experiments. It is easily checked by tracing the changes in the aliphatic and aromatic C-H bands. Because the C-H bond is hard to decompose or to form below 300°C, we can easily know whether some artifact exists or not from the variation of the 2890 and 2930 cm^{-1} bands (the absorption of aliphatic C-H bonds), and 3050 cm^{-1} band (the absorption of aromatic C-H bond). It was found that these intensities were not affected by the temperature up to 300 °C. Therefore, all the changes in the OH spectra were really what happened in coal during the heating. Now, significant changes in the OH spectra were judged to occur for MW, ND, and WY coals, whereas very little change was judged to occur for PITT and POC coals during the heating for the four coals given in Fig. 12.

4.3. Change in Strength Distribution of OH Associated with Hydrogen Bonds

All the spectra measured for the eight coals were analyzed by the procedure given above to estimate the change in HBD through the heating. **Figure 13** shows the result of peak division and the HBD distributions obtained for ND as an example. The spectra ranging from 2200 to 3650 cm^{-1} were divided into 9 peaks by a curve-fitting method. The six types of hydrogen bonds were abbreviated as follows: HB1: OH-π at 3530 cm^{-1}, HB2: OH-OH at 3400 cm^{-1}, HB3: OH-ether at 3280 cm^{-1}, HB4: cyclic OH-OH at 3150 cm^{-1}, HB5 at 2940 cm^{-1}; OH-N, and HB6: COOH-COOH at 2650 cm^{-1}. These are in the order of increasing strength. No free OH bands were detected for all the coals. In actual peak division the peak positions were slightly changed, depending on the sample by referring to the second derivative of each spectrum. The bands for the C-H bonds did not change during the heating, then they were easily separated from the hydrogen bonded bands. The amount of OH and the strength of each hydrogen bond, $(n_{OH})_j$, were calculated from each peak intensity, A_j, and the OH wavenumber shift, $\Delta\nu_{OH}$, with the aid of eq. (13). To do so, α_0 value must be estimated for each sample. If we assume that all adsorbed water is desorbed below 150 °C (the validity of this assumption is shown below), the total amount of OH (the sum of OH and COOH groups) given for each coal in Table 1 is equated to the total amount of OH at 150 °C. The total amount of OH, $(n_{OH})_{total}$, can be related to the intensity of each absorption band by

$$n_{OH,total} = \sum_j (n_{OH})_j = \frac{1}{\alpha_0}\sum_j \left[\frac{1}{\{1+0.0147(\Delta\nu_{OH})_j\}}A_j\right] \tag{15}$$

Figure 13 Six hydrogen bonded OH stretching bands and three CH bands obtained by a curve resolving method and the strength distributions of hydrogen bonds estimated by the proposed method for ND coal.

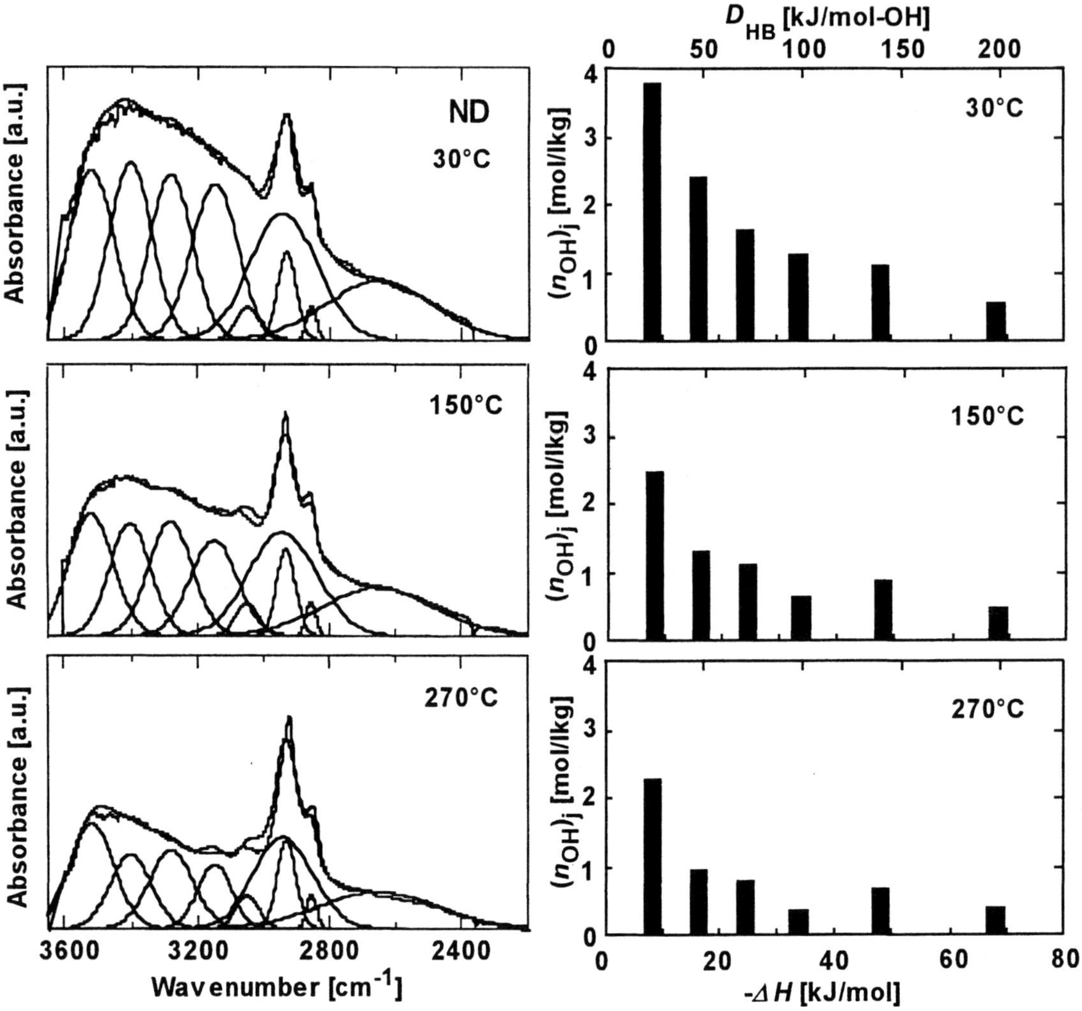

By applying this relationship to the spectrum obtained at 150 °Cα_0 value can be estimated for each coal sample. The strength distributions of hydrogen bonds thus estimated, $(n_{OH})_j$ vs. $(-\Delta H)_j$ relationships, for ND are shown in the right figure in Fig. 13. The strengths of individual hydrogen bonds represented by -ΔH values are HB1: 10 kJ/mol-OH, HB2: 17 kJ/mol-OH, HB3: 25 kJ/mol-OH, HB4: 35 kJ/mol-OH, HB5: 50 kJ/mol-OH, HB6: 70 kJ/mol-OH, and those represented by D_{HB} values are HB1: 24 kJ/mol-OH, HB2: 50 kJ/mol-OH, HB3: 74 kJ/mol-OH, HB4: 100 kJ/mol-OH, HB5: 141 kJ/mol-OH, HB6: 199 kJ/mol-OH. Since we assumed that all adsorbed water was desorbed by 150 °C, the distribution at 150 °C is the distribution of the coal dried and heated up to 150 °C. The dried ND coal is rich in rather weak hydrogen bonds: 2.5 mol-

OH/kg of HB1, 1.3 mol-OH/kg of HB2, and 1.1 mol-OH/kg of HB3. When the raw ND coal was heated from 30 °C to 150 °C, the amounts of weaker hydrogen bonds such as HB1, HB2, HB3, and HB4 decreased significantly. Since this decrease is due to the desorption of adsorbed water, it is found that water is interacting with coal hydroxyls by rather weak interactions. When the coal was further heated to 270 °C, all hydrogen bonds decreased. This decrease is judged to be due to the decomposition reactions of carboxylic groups such as eq. (14). Thus it was clarified that the *in-situ* DRIFT technique and the analysis method proposed here are powerful to examine the change in hydrogen bonds in coal.

4.4. Sensitivity of FTIR to Adsorbed Water

In the course of various spectra measurements we have experienced how FTIR is sensitive to water. When a coal sample was heated to 130 °C to remove the adsorbed water and then cooled to 30 °C in an atmosphere of flowing dry nitrogen, the absorption bands ranging from 3000 to 3500 cm^{-1} increased considerably. We have had difficulty in interpreting the unexpected change, and finally we found it is due to the adsorption of water from the "dry" nitrogen. The dried coal took up a fairly large amount of water very quickly even from the dry nitrogen containing water by less than 100 ppm. When we used an ultra high purity nitrogen containing water less than 10 ppm instead of the dry nitrogen as the carrier gas, very little change was observed between the spectrum at 130 °C and the spectrum of the coal cooled to 30 °C as typically shown for ND coal in **Figure 14.** The slight difference between the two spectra is probably due to rearrangement of hydrogen bonds upon cooling. Thus we found that *in-situ* measurements in water free atmosphere are indispensable for quantitative analysis of OH absorption bands. We were also impressed that the *in-situ* DRIFT technique would be a most sensitive method to analyze water in solid materials. These findings also suggested that the water problem cannot completely be removed when the conventional KBr pellet method was used.

4.5. Quantitative Analysis of Change in Hydrogen Bonds during the Heating

The changes of the amounts of the six hydrogen bonded OH (HB1 To HB6) were estimated by the procedure presented above, then the changes of their amounts with increasing temperature are shown for the all eight coals in **Figure 15**. For lower rank coals, MW, ND, and WY, the amounts of weaker hydrogen bonds such as HB1, HB2, HB3, and HB4 decreased significantly below 150 °C. These decreases are associated with the desorption of adsorbed water as shown for the resin in Fig. 11. With further increase of temperature the total amount of hydrogen bonds, $(n_{OH})_{total}$, decreased for MW, ND, and WY, but it was kept almost constant for IL, PITT, UF, and POC. These results show that decomposition reactions of carboxylic groups start at around 200 °C for the lower rank coals which have many COOH groups as shown in Table 1. On the other hand, such reactions are negligible for the higher rank coals, because the COOH contents are very low for these coals. It is noteworthy that only the distribution changes without affecting the total amount of hydrogen bonds for UT, PITT, and UF above 200 °C. This means that the stronger hydrogen bond, HB5, shifted to weaker hydrogen bonds, HB1 and HB2 at higher temperatures. This phenomenon seems to be associated with the glass transition.

Figure 14 *In-situ* FTIR spectrum of ND coal heated to 130 °C and the spectrum of the coal cooled to 30 °C in an flowing ultra high purity nitrogen.

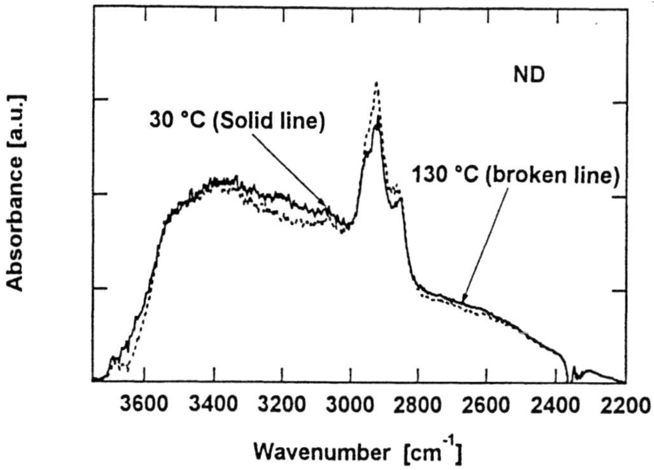

To examine the change in the hydrogen bonds more quantitatively, total enthalpy for the formation of all OH associated hydrogen bonds, $(-\Delta H)_{total}$ (< 0), at each temperature was calculated by

$$(-\Delta H)_{total} = \sum_{j} (n_{OH})_j(-\Delta H)_j \tag{16}$$

Utilizing $(-\Delta H)_{total}$ and the total amount of OH groups, $(n_{OH})_{total}$, the average strength of hydrogen bonds, $(-\Delta H)_{av}$, can also be calculated by

$$(-\Delta H)_{av} = \frac{(-\Delta H)_{total}}{(n_{OH})_{total}} = \frac{\sum_{j} (n_{OH})_j(-\Delta H)_j}{\sum_{j} (n_{OH})_j} \tag{17}$$

Inserting eqs. (2) and (13) into (17) gives

$$(-\Delta H)_{av} = \frac{\sum_{j}\left[\frac{1}{\{1+0.0147(\Delta\nu_{OH})_j\}} A_j\{0.067(\Delta\nu_{OH})_j + 2.64\}\right]}{\sum_{j}\left[\frac{1}{\{1+0.0147(\Delta\nu_{OH})_j\}} A_j\right]} \tag{18}$$

Figure 15 Change of $(n_{OH})_j$, $(n_{OH})_{total}$ and $(-\Delta H)_{total}$ with increasing temperature for the eight coals.

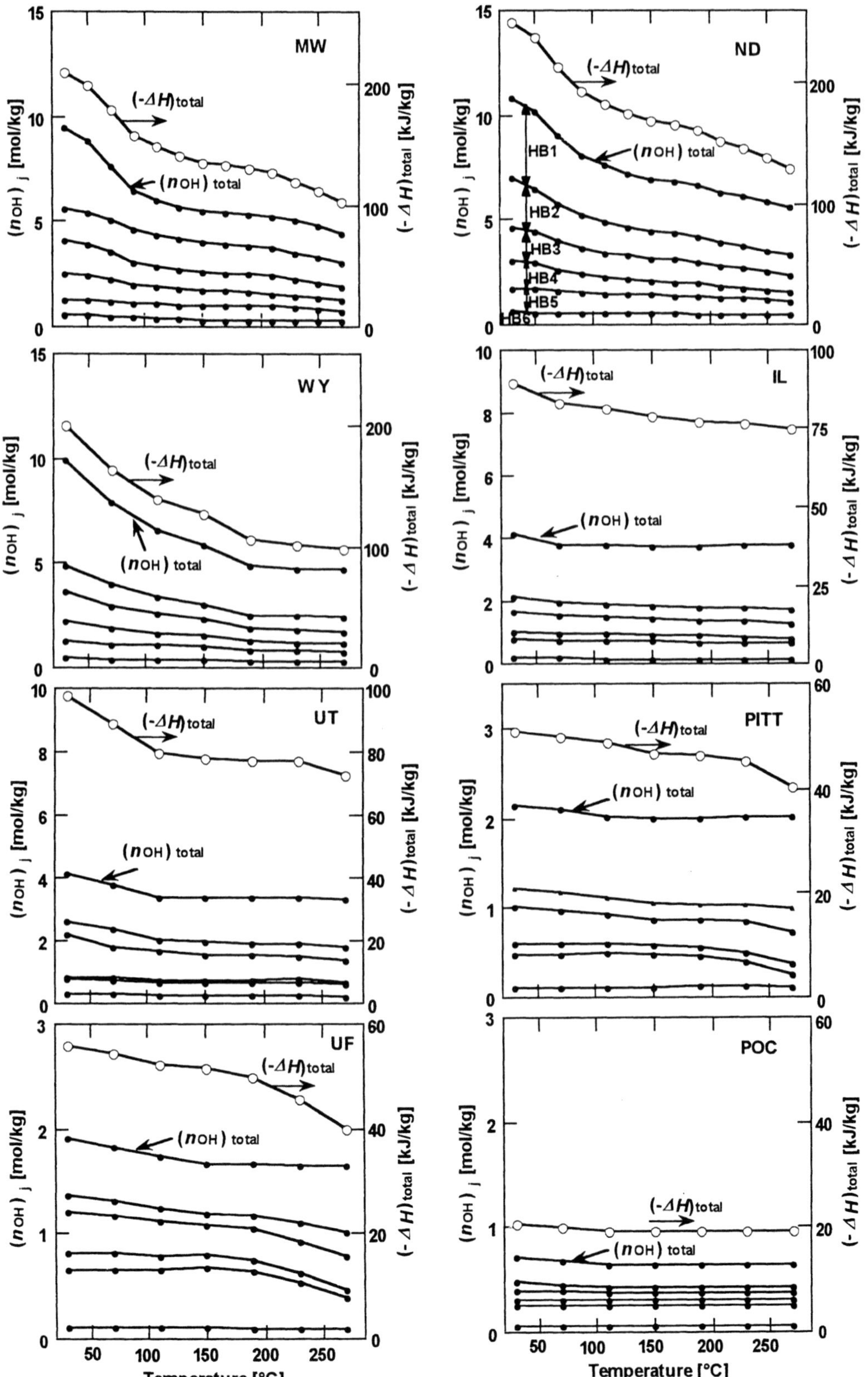

The (- $\Delta H)_{total}$ values are shown in Fig. 15, and the (- $\Delta H)_{av}$ values for the temperatures higher than 150 °C are shown in **Figure 16**. The (- $\Delta H)_{total}$ value decreases with the increase of temperature due to the change in hydrogen bonds in coal. When the number of total OH groups, $(n_{OH})_{total}$, does not change, the decrease in (- $\Delta H)_{total}$ is due to the rearrangement of hydrogen bonds. When H_2O is produced during the heat treatment, the number of OH groups decreases by the two mechanisms, and consequently (- $\Delta H)_{total}$ decreases. The first mechanism is the desorption of adsorbed water and the other is the formation of H_2O by the decomposition of COOH groups. For MW, ND and WY, both (- $\Delta H)_{total}$ and $(n_{OH})_{total}$ decreased with increasing temperature, indicting that the change in (- $\Delta H)_{total}$ is brought about by the formation of water. On the other hand, for UT, PITT, and UF, only the (- $\Delta H)_{total}$ value decreased without affecting the value of $(n_{OH})_{total}$ above 230 °C. This indicates that the rearrangement of hydrogen bonds occurred at above 230 °C as stated above. This rearrangement is well associated with the glass transition as detected by DSC by several investigators [20, 21].

Figure 16 Change in the average strength of hydrogen bonds, (- $\Delta H)_{av}$, with increasing temperature for the eight coals.

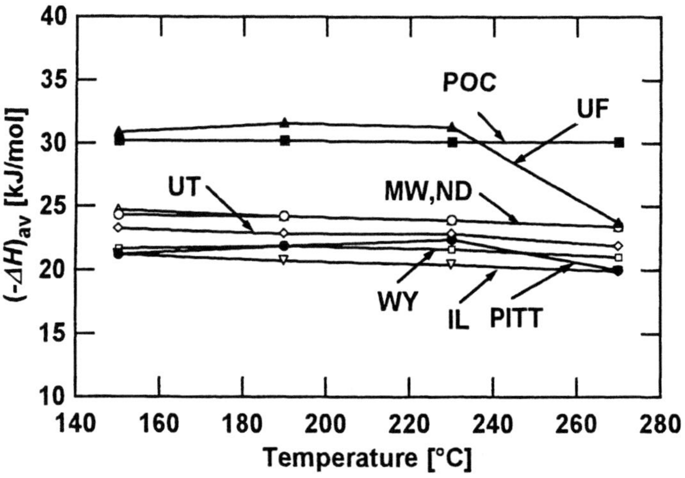

The changes in the average hydrogen bond strength, (- $\Delta H)_{av}$, in Fig. 16 are also informative. The (- $\Delta H)_{av}$ values are around 20 to 24 kJ/mol-OH and kept almost constant during the heating for MW, ND, WY, IL and UT. The (- $\Delta H)_{av}$ value for PITT is around 23 kJ/mol-OH below 230 °C, but it decreases to 20 kJ/mol-OH above 230 °C, indicating that the glass transition occurred between 230 to 270 °C. For high rank coals of UF and POC, the (- $\Delta H)_{av}$ values are 28 to 30 kJ/mol-OH, which are larger than those for the other coals by 4 to 8 kJ/mol-OH. The (- $\Delta H)_{av}$ value for UF also decreased to 23.7 kJ/mol-OH above 230 °C, again indicating the occurrence of the glass transition. The (- $\Delta H)_{av}$ values estimated here are rather close to the values reported as the strength of hydrogen bonding in the literature [1, 2, 17]. A little bit larger $(-\Delta H)_{av}$ values for UF and POC are associated with the relative abundance of OH-N hydrogen bonds (HB5) in these coals. It was clearly shown that the glass transition behavior can be detected from the change in $(-\Delta H)_{av}$ value for higher rank coals. We can now estimate the change in enthalpies through the glass transition. They are, for example, 10 kJ/kg-coal and 7 kJ/mol-OH for UF, and 6 kJ/kg-coal and 3 kJ/mol-OH for PITT.

What we would like to stress here is that the value of α_0 is not required to calculate $(-\Delta H)_{av}$ as can be found from eq. (18), which means that FTIR spectra measurements are self sufficient to obtain $(-\Delta H)_{av}$. This is one of the merits of the proposed analysis method. This facilitates the analysis, and therefore the accuracy of the $(-\Delta H)_{av}$ values is well expected.

4.6. Differentiation between Adsorbed Water and Decomposed Water

It has been rather difficult to distinguish the chemisorbed water and the water formed by decomposition reaction, which made it difficult to define the dried state of coal especially for lower rank coals. We arbitrarily assumed above that all adsorbed water desorbs by 150 °C. Now we will examine the validity of the assumption.

In this analysis we could estimate the changes in the amount of OH, $(n_{OH})_{total}$, and the enthalpy related to hydrogen bonds, $(-\Delta H)_{total}$, as shown in Fig. 15. Then we can estimate the change in the enthalpy related to hydrogen bonding with the formation of water by using the values of $(-\Delta H)_{total}$ and $(n_{OH})_{total}$ as

$$(\Delta H)_{H2O} = 2\frac{d(-\Delta H)_{total}}{d(n_{OH})_{total}} \quad [kJ/mol\text{-}H_2O] \tag{19}$$

where 2 in the r.h.s. was included to account for the fact that one H_2O molecule contains two OH groups. The magnitude of $(\Delta H)_{H2O}$ thus estimated is well expected to be dependent on the mechanism of water formation. Then the $(\Delta H)_{H2O}$ values calculated by eq. (19) with the aid of eqs. (2), (13), and (16) are shown against temperature in **Figure 17.** The values obtained for the resin and the lignin are also included in the figure for comparison. The $(\Delta H)_{H2O}$ vs. temperature relationships were so close each other, and there was a distinct jump at about 140 °C: the $(\Delta H)_{H2O}$ values below 140 °C were close to or slightly smaller than the heat of vaporization of water, ΔH_{vap}, whereas the $(\Delta H)_{H2O}$ values above 140 °C were significantly larger than ΔH_{vap}. These results suggest that the water detected below 140 °C is the adsorbed water and that formed above 140 °C is the water formed by some decomposition reaction. Thus the HBD estimated by the proposed method is found to be utilized to distinguish the type of water formed. We would like to stress again that the value of a_0 or the value of $(n_{OH})_{total}$ is not required to obtain $(\Delta H)_{H2O}$, as was the case to calculate $(-\Delta H)_{av}$. Only FTIR spectra measurements are enough to obtain $(\Delta H)_{H2O}$.

Figure 17 Changes in the water formation enthalpy, $(\Delta H)_{H2O}$, with increasing temperature for the coals, the resin, and the lignin.

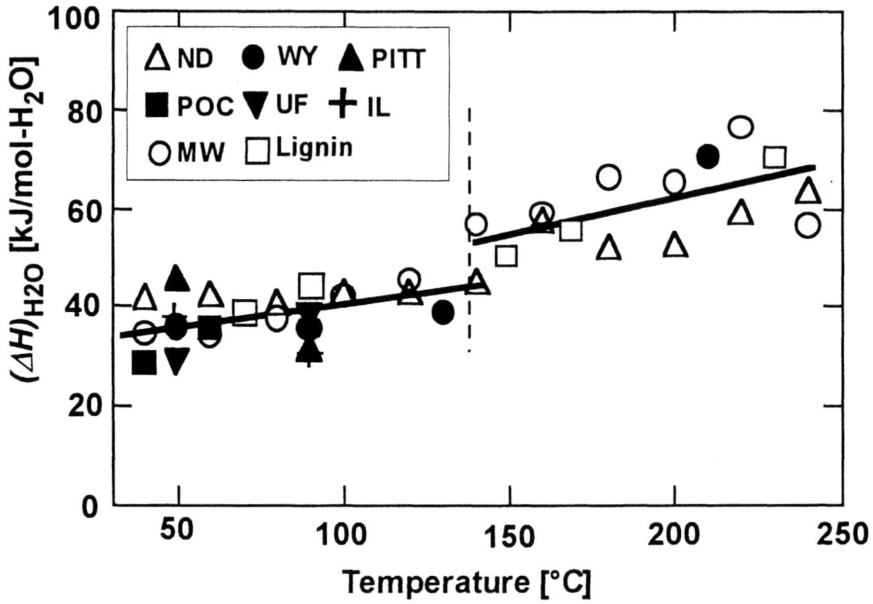

5. Conclusion

A new method was presented to estimate the strength distribution of hydrogen bonds in coal, and its validity was examined by analyzing the change in the hydrogen bond distributions of 8 different coals, a resin, and a lignin during their heat treatment below 300 °C. The hydrogen bonds examined included the coal intramolecule hydrogen bonds and coal-water hydrogen bonds formed by hydroxyls in coal. The method analyzes the FTIR spectrum ranging from 2400 to 3700 cm^{-1} that is obtained using the *in-situ* diffuse reflectance IR Fourier transform (DRIFT) technique with neat, undiluted, coal samples.

The analysis method and the experimental technique employed clarified the following.

1. Water is taken up on coal even from a dry atmosphere so quickly that *in-situ* measurements in a completely water free atmosphere are indispensable for quantitative analysis of OH absorption bands. This suggested that the water problem cannot completely be removed when the conventional KBr pellet method was used.
2. Coal itself has hydrogen bonds ranging from 10 kJ/mol to 70 kJ/mol in strength. The relative abundance of the hydrogen bonds is highly dependent on coal type. Lower rank coals are rich in weaker hydrogen bonds associated with OH-π, OH-OH, and OH-ether, whereas some high rank coals are rich in OH-N hydrogen bonds.
3. The average strength of hydrogen bonds was found to decrease above 230 °C for some bituminous coals without affecting the total number of hydrogen bonds.

This clarified that the change accompanying the glass transition can be well traced by the proposed analysis method.

4. The enthalpy associated with the formation of water could be successfully estimated solely from the FTIR measurement. The enthalpy increased gradually up to 140 °C from 30 kJ/mol to 40 kJ/mol, but it increased stepwisely to a level of 50 kJ/mol over 140 °C. This clarified that the water formation below 140 °C is due to the desorption of adsorbed water, and that the water formation above 140 °C is due to the decomposition reactions of COOH groups. This finding enabled a definition of the dried state of coal. It is also possible to quantitatively examine the early stage of pyrolysis of coal based on this information.

Thus the validity and usefulness of the proposed method was clarified. We are sure that this method will be extended to analyze various solvent-coal interactions. The extension of this method to other fields may also be well expected.

ACKNOWLEDGEMENT

This work was supported by "Research for the Future" project of the Japan Society for the Promotion of Science (JSPS) through the 148 Committee on Coal Utilization Technology of JSPS. Contributions of Dr. Wen Li, Messrs. F. Morozumi and T. Kusakawa, and Ms. A. Kumano to this work are gratefully acknowledged.

REFERENCES

[1] Larsen, J. W., Green, T. K., and Kovac, J., *J. Org. Chem*, **50**, pp. 4729-4735, (1985)

[2] Larsen, J. W. and Shawver, S., *Energy &Fuels*, **4**, pp. 74-77, (1990)

[3] Lucht, L. M. and Peppas, N. A., *Fuel*, **66**, pp. 815-826, (1987)

[4] Lucht, L. M. and Peppas, N. A., *J. Appl. Polym. Sci.*, **33**, pp. 2777-2785, (1987)

[5] Suuberg, E. M., Otake, Y., and Sezen, Y., DOE Report "Diffusion in Coals", (DOE/PC/80527-12), (1990)

[6] Aida, T. and Squires, T. G., *Prepr.Pap.- Am. Chem. Soc. Fuel Chem.*, **30**, pp. 95-102, (1985)

[7] Miura, K., Mae, K., Takebe, S., Wakiyasu, H., and Hashimoto, K., *Energy &Fuels*, **8**, pp. 874-880, (1994)

[8] Mae, K., Maki, T., Okutsu, H., and Miura, K., *Fuel*, **79**, pp. 417-425, (2000)

[9] Sanada, Y. and Honda, H., *Fuel*, **45**, pp. 295-300, (1966)

[10] Larsen, J. W. and Gurevich, I., *Energy &Fuels*, **10**, pp. 1269-1272, (1996)

[11] Lucht, L. M., Larson, J. M., and Peppas, N. A., *Energy &Fuels*, **1**, pp. 6-58, (1987)

[12] Mackinnon, A. J. and Hall, P. J., *Energy &Fuels*, **9**, pp. 25-32, (1995)

[13] Yun, Y. and Suuberg, E. M., *Fuel*, **72**, pp. 1245-1254, (1993)

[14] Iyengar, M. S. and Lahiri, A., *Fuel*, **36**, pp. 286-297, (1957)

[15] Pauling, L., "*The Nature of the Chemical Bond*", Cornell University Press, Ithaca, N.Y., p. 284, (1948)
[16] Allardice, D. J. and Evans, D. G., *Fuel*, **50**, pp. 236-253, (1970)
[17] Glass A. S. and Larsen, J. W., *Energy &Fuels*, **8**, pp. 284-285, (1994)
[18] Arnett, E. M., Mitchel, E. J., Murty, T. S. S. R., Gorrie, T. M., and Schleyer, P. V. R., *JACS*, **92**, pp. 2365-2377, (1970)
[19] Solomon, P. R., "Relation between Coal Structure and Thermal Decomposition Products", in "Coal Dstructure: Advances in Chemistry Series", Gorbaty, M. L. and Ouchim, K., Eds., ACS, Washington, D. C., vol. 192, Chap. 7, p. 95, (1981).
[20] Solomon, P. R., Hamblen, D. G., and Carangelo, R. M., *Application of Fourier Transform IR Spectroscopy in Fuel Science, in "Coal and Coal Products: Analytical Characterization Technique; Symposium Sereies*", Fuller, E. L., Eds., ACS, Washington, D. C., vol. 205, Chap. 4, p. 77(1982).
[21] Painter, P. C., Snyder, R. W., Starsinic, M., Coleman, M. M., Kuehn, D. W., and Davis, A., "*Fourier Transform IR Spectroscopy – Aplication to the Quantitative Determnination of Functional Groups*", Chap. 3, p. 47, *idem.*
[22] Smyrl, N. R. and Fuller, E. L. Jr., "*Chemistry and Structure of Coals – Diffuse Reflectance IR Fourier Transform (DRIFT) Spectroscopy of Air Oxidation*" Chap. 5, p. 133, *idem.*
[23] Solomon, P. R. and Carangelo, R. M., *Fuel*, **61**, pp. 663-669, (1982)
[24] Snyder, R. W., Painter, P. C., Havens, J. R., and Koeniug, J. L., *Applied Spectroscopy*, **37**, pp. 497-502, (1983)
[25] Fuller, E. L. Jr. and Smyrl, N., *Fuel*, **64**, pp. 1143-1150, (1984)
[26] Painter, P. C., Sobkobiak, M., and Youtcheff, J., *Fuel*, **66**, pp. 973-978, (1987)
[27] Mu, R. and Malhotra, V. M., *Fuel*, **70**, pp. 1233-1234, (1991)
[28] Sobkobiak, M. and Painter, P. C., *Energy &Fuels*, **9**, pp. 359-363, (1995)
[29] Chen, C., Gao, J., and Yan, Y., *Energy &Fuels*, **12**, pp. 446-449, (1998)
[30] Solomon, P. R. and Carangelo, R. M., *Fuel*, **67**, pp. 949-959, (1988)
[31] Larsen, J. W. and Basker, A. J., *Energy &Fuels*, **1**, pp. 230-232, (1987)
[32] "*Hydrogen Bonding*", Hadzi, D. and Thompson, H. W., Eds., Pergamon Press Ltd., London, (1959)
[33] "*Hydrogen Bonding*", Vinogradov, S. N. and Linnell, R. H., Eds., Van Nostrand Reinhold Co., N.Y., (1971)
[34] "*The Hydrogen Bond, Recent Developments in Theory and Experiments*", Schuster, P., Zundel, G., and Sandorfy, C., Eds., North Holland Publishing Co., N.Y., Oxford, (1976)
[35] Joesten, M. D. and Drago, R. S., *JACS*, **84**, pp. 3817-3821, (1962)
[36] Purcell, K. F. and Drago, R. S., *JACS*, **89**, pp. 2874-2879, (1967)
[37] Epley, T. D. and Drago, R. S., *JACS*, **89**, pp. 5770-5773, (1967)
[38] Drago, R. S. and Epley, T. D., *JACS*, **91**, pp. 2883-2890, (1969)
[39] Arnett, E. M., Mitchel, E. J., and Murty, T. S. S. R., *JACS*, **96**, pp. 3875-3891, (1974)
[40] Sherry, A. D., *Spectroscopic and Calorimetric Studies of Hydrogen Bonding,* in Lit. **34**.

[41] Miura, K., Mae, K., and Morozumi, F., *Prepr.Pap.- Am. Chem. Soc. Fuel Chem.*, **42**(1), pp. 209-213, (1997)
[42] Miura, K., Mae, K., Li, W., Kusakawa, T., and Kumano, A., *Prepr.Pap.- Am. Chem. Soc. Fuel Chem.*, **44**(3), pp. 642-646, (1999)
[43] Detoni, S., Hadzi, D., and Smerkolj, R., *J. Chem. Soc. (A)*, pp. 2851-2859, (1970)
[44] Kiselev, A. V., *Surface Science*, **3**, pp. 292-293, (1965)
[45] Jung, B., Stachel, S., and Calkins, W. H., Preprint, *Am. Chem. Soc., Div. Fuel Chem.*, **36**(3), pp. 869-876, (1991)
[46] Schafer, H. N. S., "*Functional Groups and Ion Exchange Properties*", in "*The Science of Victorian Brown Coal*", Durie, R. A., Ed. Butterworth Heinemann, Chap. 7, p. 334, (1991)

Chapter 2

REMOVAL OF COAL SULFUR DUE TO RAPID PYROLYSIS WITH CHEMICAL PRETREATMENTS

***Katsuyasu Sugawara**, *Takuo Sugawara** and *Masayuki Shirai*†**

1. INTRODUCTION

The formation of CO_2, SO_X, NO_X and ash has been unavoidable in conventional coal utilization processes. Development of precleaning technology for removal of sulfur, nitrogen and ash-forming minerals as well as advancing coal utilization process with high efficiency is desirable in order to minimize the effluents of environment-unfriendly wastes and to maintain fossil energy resources in limited reserves.

Coal pyrolysis occurs invariably in any coal conversion process due to heating, and volatile matter sometimes evolves more than 50% of raw coal. Sulfur is a key component not only for environmental conservation but also for the development of new coal utilization processes such as A-PFBC and IGFC etc. [1] Incomplete understanding of the behavior of organic sulfur during heat treatment has given rise to the common view that organic sulfur in coal cannot be removed by pretreatments.

Conventional coal cleaning technologies are mainly grouped into physical and chemical methods which aim to remove inorganic sulfur and coarse mineral matter, while chemical methods attempt to lower the content of organic sulfur and fine mineral matter distributed in the coal matrix. Because the physical process is limited to removal of inorganic constituents, development of an efficient process of organic sulfur removal is an urgent issue for sustainable resource development and environmental conservation.

Biological treatments can be used for pyritic sulfur removal and bacteria that effectively remove organic sulfur have been recently found [2]. Molten caustic leaching [3], wet oxidation by potassium permanganate [4], and microwave irradiation [5] have also been reported as chemical cleaning methods. Of the many chemical desulfurization

* Faculty of Engineering and Resource Science, Akita University, 1-1 Gakuen-cho, Akita City, Akita Prefecture 010-8502, Japan (katsu@ipc.akita-u.ac.jp).
† Institute for Chemical Reaction Science, Tohoku University, Katahira, Aoba-ku, Sendai, 9802-8577 Japan.

methods reported, only TRW Gravimelt and CSIRO molten caustic leaching processes have been conducted on a pilot plant scale. In these processes sodium hydroxide or potassium hydroxide and coal are mixed and heated to 400 °C [6]. Potassium hydroxide has been reported to be a more effective leachant than sodium hydroxide for high-rank coal [3].

Development of a desulfurization process requires accurate understanding of organic sulfur forms in coal. Recently, X-ray absorption near-edge structure (XANES) and X-ray photoelectron spectroscopy (XPS) have been demonstrated as strong tools to elucidate organic sulfur forms in coal and heavy hydrocarbons [7, 8, 9, 10]. These analytical results indicate that aliphatic sulfur decreases inversely to aromatic sulfur with the increase of carbon content in coal.

A few investigations have been reported on dynamic behavior of sulfur forms as well as total sulfur change in pyrolysis of coal while many desulfurization studies were previously carried out only for total sulfur reduction. Precise knowledge and kinetic evaluation of the dynamic behavior of sulfur forms under various heat-treatment conditions are necessary for the development and design of chemical coal cleaning processes.

In the present study, a desulfurization scheme of coal was proposed to express the change in sulfur forms (pyritic, ferrous sulfide, sulfate, sulfite and organic sulfur) during pyrolysis in different heating rates and temperature with the use of fixed bed and free-fall reactor. Density separated coals were pyrolyzed rapidly to investigate the effects of sulfur forms and their distribution, and aromaticity on desulfurization characteristics. Combined processes of some chemical pretreatments with rapid pyrolysis are introduced which has aimed to transform the thermally stable form of organic sulfur to reactive species before pyrolysis. Changes in organic sulfur forms were determined by using XANES for coal and chars.

2. Sulfur Behavior during Pyrolysis

2.1. Effect of Heating Rate and Temperature

Many studies on desulfurization of coal indicated that the removal of sulfur from solid phase depends on the treatment temperature, gas composition, ambient pressure and kind of coal. These results were, however, only focused on the total sulfur change in the char, including the influence of experimental conditions and the difference of apparatus, and did no more than provide a qualitative understanding of total sulfur behavior without kinetic interpretation.

Some researchers have reported the existence of optimum desulfurization temperature in the pyrolysis of some coals [11, 12]. The present authors also reported a unique observation that organic sulfur change varied through a minimum during pyrolysis of coal containing abundant pyritic sulfur [13]. This behavior of organic sulfur was interpreted as a result of interaction between organic and inorganic sulfurs depending on coal type. Knowledge of the optimum desulfurization temperature is important not only to understand the desulfurization mechanism but also for the design of pyrolysis desulfurization processes.

A noncaking bituminous coal (Ermelo, South Africa) was pyrolyzed using a fixed bed in a nitrogen stream at an atmospheric pressure up to 900 °C with heating rates of 20 – 100 °C/min [14]. Figure 1 shows the change in organic sulfur content of the solid phase with final pyrolysis temperature for the two different heating rates. The sample coal was heated linearly to the desired temperatures at 300 – 900 °C with heating rates of 20 and 100 °C/min, and then cooled down with a 200 °C/min average cooling rate without holding at the final temperature. For 20 °C/min heating rate, the organic sulfur content decreased from 0.48 % of raw coal to the minimum of 0.3 % for char obtained at 500 °C and then increased to 0.48 % for char obtained at 900 °C. Under the condition of a 100 °C/min heating rate, the change in organic sulfur content also shows a minimum of 0.27 % obtained at 500 °C. The present results indicate that heating rate as well as final pyrolysis temperature plays an important role in desulfurization in that the sulfur content in char was found to be lower at a 100 °C/min than at a 20 °C/min heating rate.

Figure 1 Change in organic sulfur content in solid phase with terminal pyrolysis temperature in different heating rates.

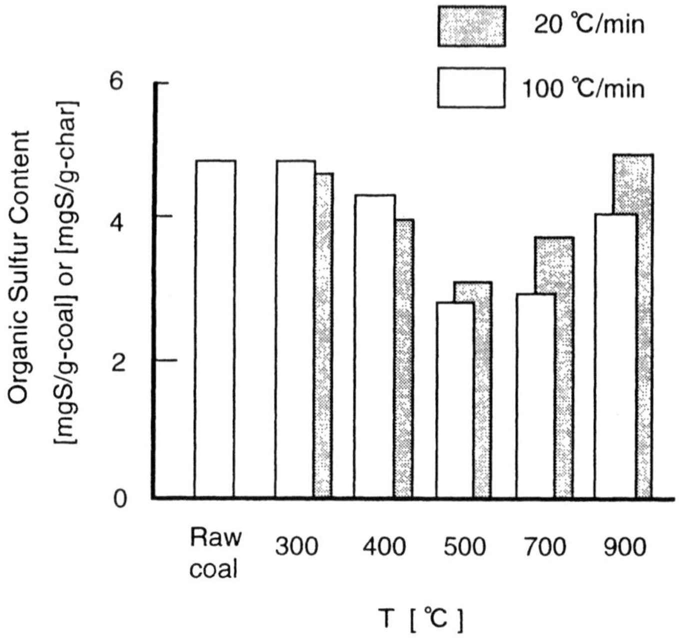

Figure 2 represents a desulfurization scheme for a coal particle. Organic sulfur in raw coal changes, partly into refractory, i.e. thermally stable, organic sulfur in the char, partly to tar, and partly into hydrogen sulfide. Pyritic sulfur forms ferrous-sulfide sulfur and hydrogen sulfide. The reversible reaction between ferrous-sulfide sulfur and metallic iron occurs under pressurized hydropyrolysis. A part of hydrogen sulfide produced by the above reactions of organic and inorganic forms is captured reversibly as a form of refractory organic sulfur in char. Hydrogen sulfide and organic sulfur are emitted from the particle surface through intrapores as parts of the volatile matter. The residence time of hydrogen sulfide within a particle in the temperature range from 300 to 900 °C can be estimated from the difference of volatile release rate to be one fourth to one half less at 100 °C/min than at 20 °C/min heating rate. The rapid release rate of volatile matter at high heating rate indicates how a low sulfur char can be obtained by controlling the competing phenomena of the uptake reaction and the diffusion in a coal particle.

Figure 2 Desulfurization scheme of coal.

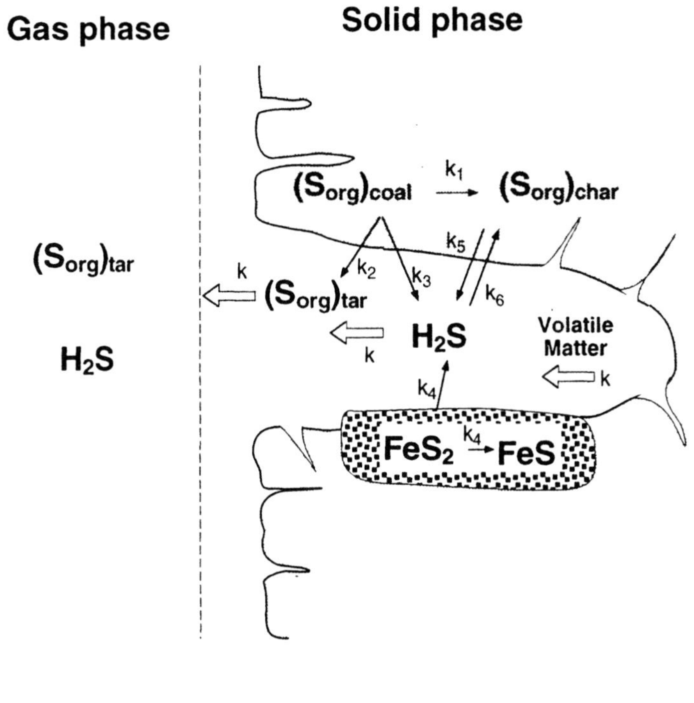

Figure 3 indicates changes in organic sulfur content in the solid phase with reaction time under the condition of 6000 °C/s heating rate and a final temperature of 960 °C by using a free fall reactor, which was a fused silica tube (i.d. 36 mm), the temperature of which was controlled by an electric furnace composed of five sections, each 30 cm long and controlled independently. Particle residence time was varied by changing the length of the heating section. The details of the apparatus were described elsewhere [15, 16]. The estimated intraparticle diffusion time of 10^{-2} sec is three orders smaller than that of slow heating experiments. Under the rapid pyrolysis condition, the removal of organic sulfur progressed with elapse of reaction time without any increase of refractory organic sulfur. The rapid pyrolysis is effective for removal of the organic sulfur in a very short reaction time with negligible uptake reaction of hydrogen sulfide.

Figure 3 Change in organic sulfur content in solid phase with reaction time in rapid pyrolysis.

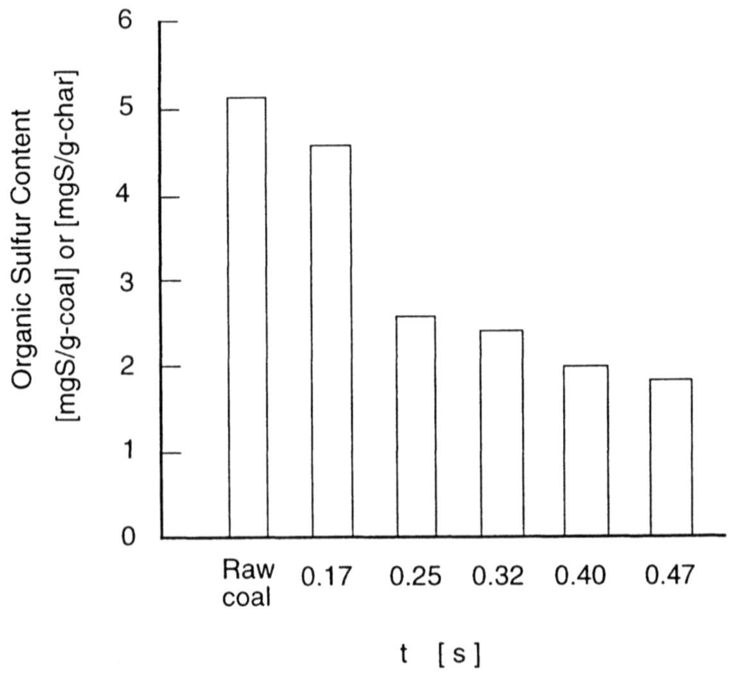

2.2. Effect of Coal Type on Desulfurization

Figure 4 shows a correlation between the extent of organic sulfur removal and carbon content for 17 coals ranging from 65 to 90 % daf. The extent of organic sulfur removal is defined by {1 – (organic sulfur in char) / (organic sulfur in raw coal)} x 100 [%] where the ultimate content of organic sulfur in the rapid pyrolyzed char based on coal basis was used as organic sulfur in char in this equation. George et al. found a linear increasing tendency of the percentage of thiophenic sulfur to total organic sulfur from 45 to 95 % with the increase of carbon content [17]. Kelemen et al. reported that thermal reactions during carbonization caused the increase of aromatic sulfur in high-rank coals since aromatic sulfur increased with treatment temperature comparing to the sulfur forms in raw coal and chars [8]. Huffmann et al. also reported that the content of sulfide sulfur decreased with the increase in carbon content [10]. By considering the thermal stability of these organic sulfur forms, it is plausible that the extent of organic sulfur removal decreases linearly with the increase of carbon content due to the increase of thiophenic sulfur content.

Figure 4 Change in extent of organic sulfur removal from solid phase with carbon content.

a. Fushun	i. PSOC-1493
b. PSOC-830	j. PSOC-592
c. Fuxin	k. Muswellbrook
d. Obed Mountain	l. Furong
e. Ermelo	m. Yanzhou (B)
f. Rosebud	n. Nantong
g. Sheerness	o. Zhong Liang Shan
h. Smoky River	p. Yanzhou (X)
	q. Mequinenza

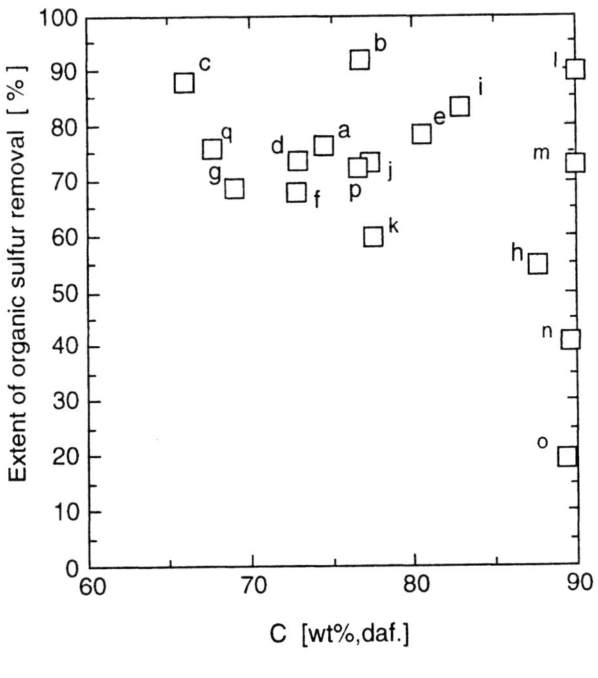

An apparent linear correlation, however, cannot be obtained between the extent of organic sulfur removal in rapid pyrolysis and carbon content of the raw coals as demonstrated in Figure 4 [18]. The samples containing around 77 % carbon show the wide spread of the extent of organic sulfur removal irrespective of the similar carbon content. For example more than 30% of the difference was observed for the extent of organic sulfur removal between PSOC 830 and Muswellbrook coals. Moreover, the large difference from 19 % to 89 % of the extent of organic sulfur removal was noted for five kinds of high rank coals containing more than 87 % carbon.

Dependency of the extent of organic sulfur removal was investigated on aromatic carbon fraction f_a of parent coal defined and calculated by the following equation [19]:

$$f_a = \{1200 \times (100 - \text{Volatile matter})\} / (1240 \times \text{Carbon } \%)$$

The aromatic carbon fraction is evaluated from several methods based on density, volatile matter, heat of combustion and solubility etc. The volatile matter based evaluation was chosen for the present study because volatile matter yields is one of important indexes of coal reactivity in pyrolysis. Figure 5 shows the relation between the extent of organic sulfur removal and aromatic carbon fraction. SS002–SS017 are standard samples supplied from the Basic research Associated for Innovated Coal Utilization Program supported by New Energy and Industrial Technology Development Organization in Japan. SS017 having the smallest value of aromatic carbon fraction shows the lowest extent of organic sulfur removal. Furthermore, SS004 was desulfurized most effectively to 90 % though the medium value of aromatic carbon fraction was obtained in the samples. An apparent linear correlation cannot be observed between the extent of organic sulfur removal and the aromatic carbon fraction. The extent of nitrogen removal, however, decreases linearly with an increase of the aromatic carbon fraction.

Figure 5 Dependency of extents of organic sulfur and nitrogen removal on aromatic carbon fraction.

a. SS017 f. SS004 k. Muswellbrook
b. SS006 g. SS010 l. Illinois No.6
c. SS002 h. SS009 m. Nantong
d. SS012 i. SS008
e. SS007 j. Mequinenza

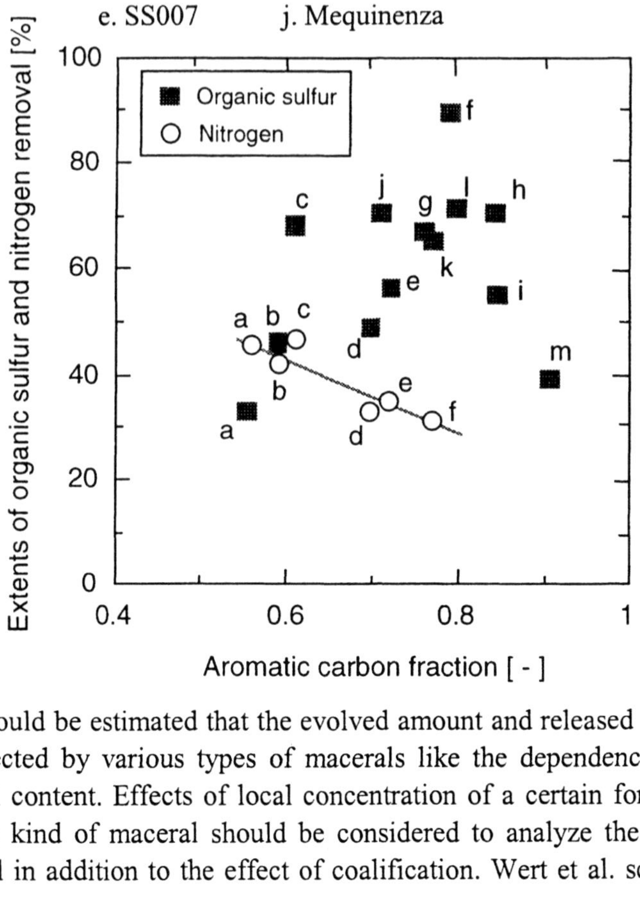

It could be estimated that the evolved amount and released rate of sulfur compounds are affected by various types of macerals like the dependency of volatiles release on maceral content. Effects of local concentration of a certain form of organic sulfur in a specific kind of maceral should be considered to analyze the extent of organic sulfur removal in addition to the effect of coalification. Wert et al. scanned the organic sulfur distribution in PSOC 592 by using EPMA and found a high sulfur content of 4.5 % in sporinite against 2 % in vitrinite [20].

2.3. Effect of Particle Density

Figure 6 shows a relation between yield and density obtained by the sink-float method in zinc chloride aqueous solution. The density distribution is different from coal to coal even though carbon content is the same. The samples were separated into three groups I to III with expected differential concentration of macerals. The range of each group was altered in four kinds of coal to obtain enough amount of fractionated samples to be supplied for rapid pyrolysis experiments.

Figure 6 Yield distribution in density separation.

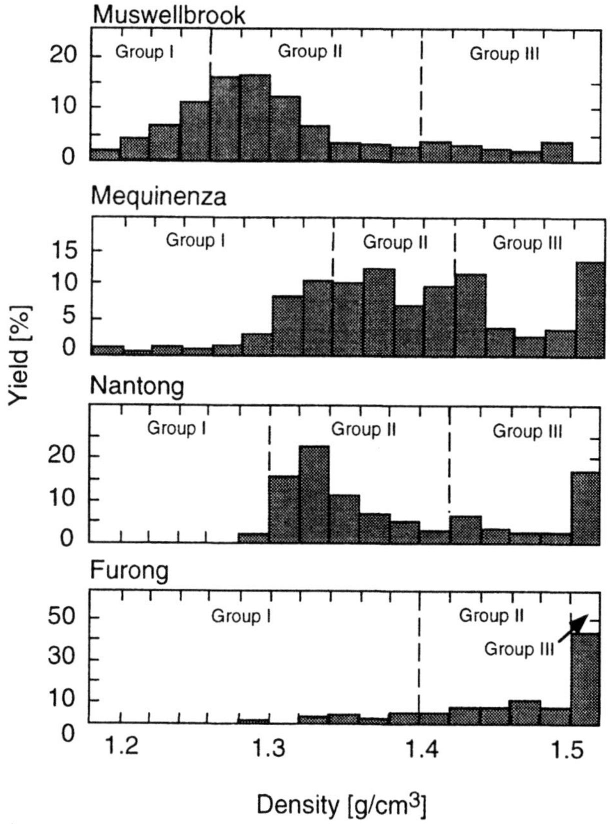

Sequential changes in the sulfur-form distribution for groups I and III of Muswellbrook coal are shown in Figure 7 [21]. The organic sulfur content decreases rapidly within 0.38 s and attains an equilibrium value. Reactions liberating organic sulfur cause efficient desulfurization, 76 %, the desulfurization extent of organic sulfur. Half the initial organic sulfur is released to tar during rapid pyrolysis of group I. For group III, the reduction of pyrite to ferrous sulfide was completed within 0.3 sec. Although organic sulfur was also reduced in a short reaction time, the desulfurization extent of organic sulfur was only 34 %, and the smallest amount of sulfur was transferred to tar from solid among the three groups.

Figure 7 Change in sulfur-form distribution for density-separated coal. (Muswellbrook; (a) light fraction, (b) heavy fraction)

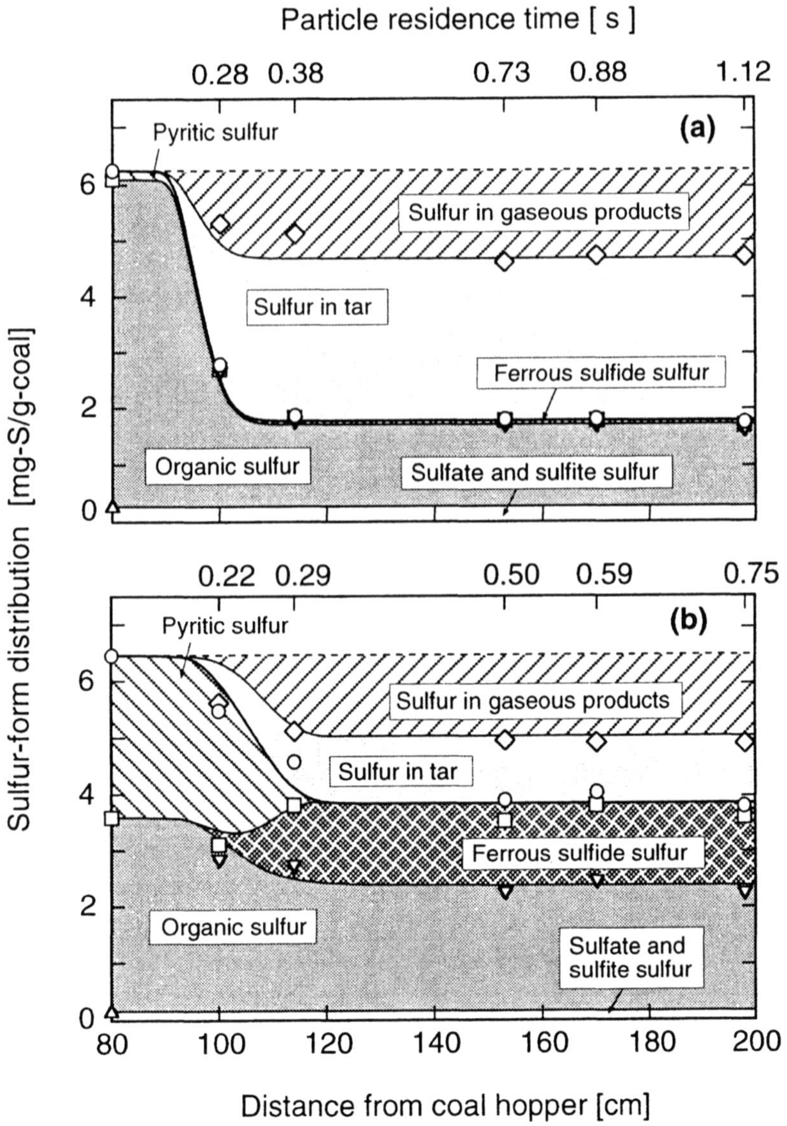

Figure 8 represents a relation between the extent of organic sulfur removal and average density. Mequinenza, Muswellbrook, and Nantong coals show linear relationship between the extent of organic sulfur removal and average density of the groups in each coal. Large extent of organic sulfur removal was observed for lower rank coal in these three coals. Higher rank coal provides steeper slope in this figure. These results imply that thermally stable organic sulfur increases with the increase on density and carbon content [18].

Figure 8 Change in organic sulfur removal with average density.
O, Group I Δ, Group II □, Group III

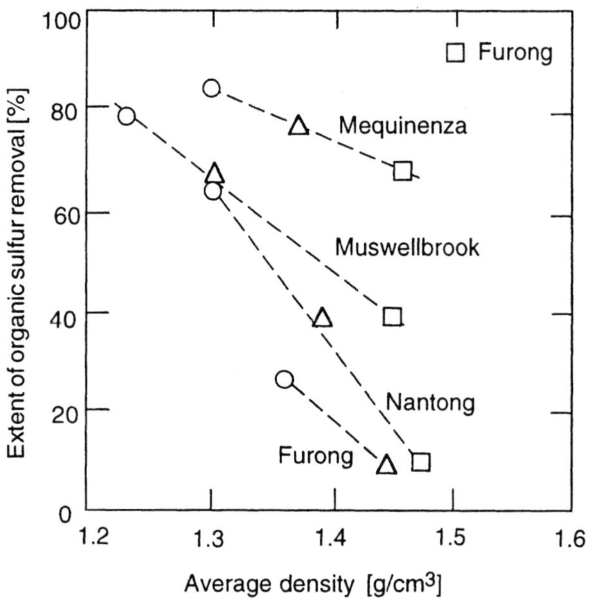

Sulfur K-edge X-ray adsorption near edge structure spectroscopy was applied to specifying the forms of organic sulfur in raw coal and char. XANES measurements were carried out at beam line 2A of Photon Factory in the National Laboratory for High Energy Physiscs [18, 22]. Figure 9 shows third differential curves of XANES spectra for groups I and III of Nantong coal. Strong peak is observed around 2472 eV for group I, which are attributable to thiophene and sulfide. A strong peak at 2474.5 eV results from sulfoxide. Though group III shows the similar distinct peaks of thiophene and sulfoxide to group I, the sulfide peak does not appear for group III. It could be estimated that efficient desulfurization in a smaller density group such as group I is related to the presence of sulfide though the absolute amount of each sulfur from should be determined. Third differential curves of XANES spectra for groups I and III of Furong coal are shown in Figure 10. Thiophene, sulfoxide, and pyritic sulfur are observed for group I as the same as the curves of Nantong, group I. In addition to the peaks resulting from thiophene, sulfoxide, and sulfate, intensive peaks are observed at 2470.4 – 2470.8 eV, which are combined peaks attributable to pyritic sulfur and elemental sulfur.

Figure 9 Sulfur XANES spectra of groups I and III for Nantong coal.

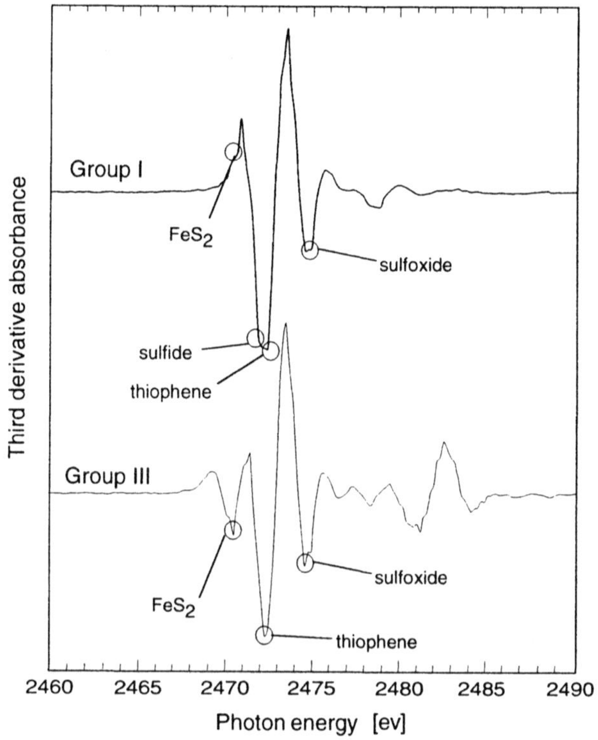

Figure 10 Sulfur XANES spectra of groups I and III for Furong coal.

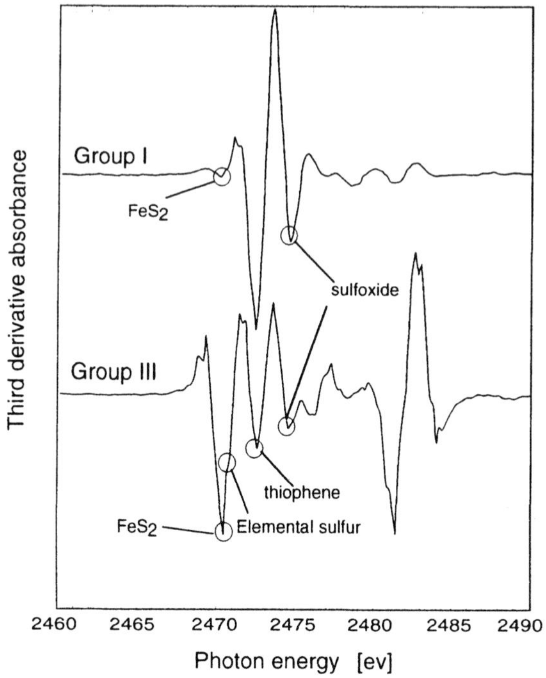

Figure 11 represents third differential curves of XANES spectra of rapid pyrolysis chars for groups I and III of Furong coal. The curve of char for group I indicates that the sulfoxide peak becomes smaller after pyrolysis while the peak of thiophenic sulfur remained. A peal at 2468.7 eV resulted from ferrous sulfide produced by reduction of pyrite. The peaks of pyritic sulfur and elemental sulfur disappeared and a large peak of ferrous sulfide at 2468.7 eV newly appeared after pyrolysis for group III. Furong coal has distinguishing characteristics as high organic sulfur content and removal extent of organic sulfur in large density group III. XANES analysis, however, indicated that efficient extent of organic sulfur removal might be caused by releasing of elemental sulfur rather than organic sulfur.

Figure 11 Sulfur XANES spectra of Nantong coal.

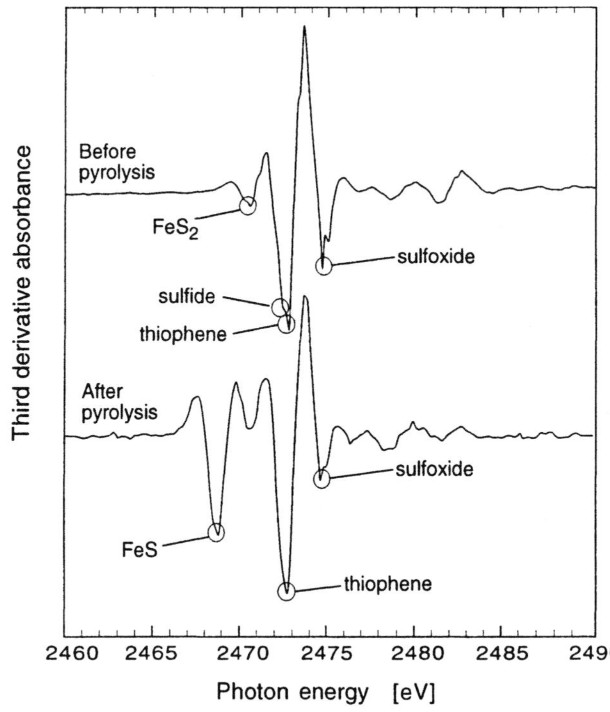

While an apparent correlation could not be found between the extent of organic sulfur removal and aromatic carbon fraction of parent coal in Fig. 5, this correlation has been tried for the density-separated samples. The new correlation is shown in Figure 12. The range of aromatic carbon fraction becomes wider than that of parent coal. The extents of organic sulfur removal decrease smoothly with the increase of aromatic carbon fraction. The four kinds of coal from lignite to high volatile bituminous showing high extent of organic sulfur removal, are grouped with the identicaldependency of organic sulfur release on the aromaticity of coal matrix. A high rank coal, Nantong indicates that a larger quantity of thermally stable organic sulfur exists even if the aromatic carbon fraction is the same.

Figure 12 Relation between extent of organic sulfur removal and aromatic carbon fraction of density-separated coals.

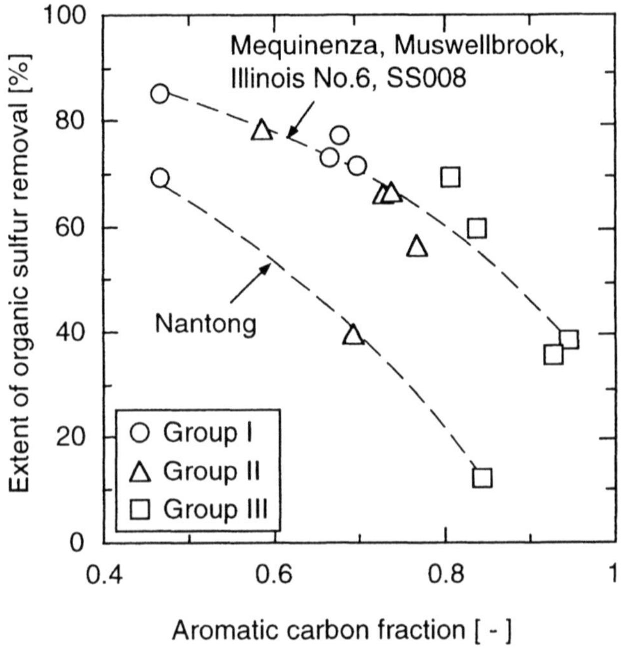

3. Rapid Pyrolysis with Chemical Pretreatments

3.1. Alkali Pretreatment

The rapid pyrolysis effectively removed organic sulfur from the solid phase in some types of coal. Part of the organic sulfur exists in thermally stable form that is difficult to remove from the solid phase by pyrolysis. Thermally stable organic sulfur remaining in the solid should be changed to decomposable forms to achieve efficient desulfurization in the pretreatment process. In the present study, three kinds of subbituminous and bituminous coals were treated with potassium hydroxide solution under milder conditions than those of Gravimelt and CSIRO processes.

Figure 13 shows an experimental procedure [23]. The coal samples were immersed into a saturated aqueous solution of potassium hydroxide for 24 h with agitation at room temperature. The wet alkali-immersed samples were then heated at 250 °C for 4 h in a nitrogen atmosphere after filtration from potassium hydroxide solution. The alkali-treated samples were washed repeatedly with distilled water to reduce the basicity and then dried at 107 °C. The parent coals and their alkali-treated products were pyrolyzed rapidly in a nitrogen atmosphere in a free-fall reactor.

Figure 13 Experimental procedure of alkali pretreatment.

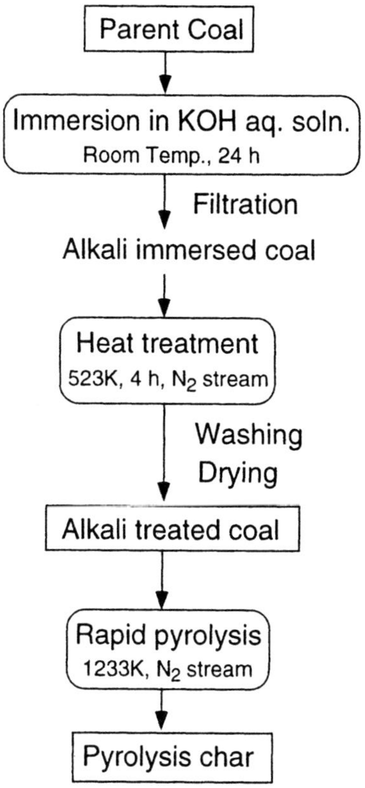

Figure 14 shows sequential changes in the distribution of the forms of sulfur for Muswellbook and its alkali-treated products versus the distance from the coal hopper or the residence time of the particle in the reactor. For the parent coal shown in Fig. 14(a), the total sulfur content in the solid phase (indicated by circles) deceased rapidly from 0.66 to 0.29 % within 0.33 sec and attained a constant value corresponding to the release of volatile matter. The ultimate extent of organic sulfur removal was 60 % in rapid pyrolysis of the parent coal. The alkali-treated sample (Fig. 14(b)) showed a rapid decrease of sulfur and 94 % ultimate organic sulfur removal. Hence the combined process of rapid pyrolysis and alkali treatment was very successful in reducing organic sulfur in Muswellbrook coal. Thiophenic sulfur is a major component of the organic sulfur in coal, its proportion increasing from 55 to > 90 % with increasing carbon content from 75 to 92 % daf for Argonne Premium coals. Calkins reported that thiophenic compounds showed a low conversion of < 3 % at 950 °C whereas aliphatic and aromatic sulfides, mercaptans and disulfide decomposed with high conversion at < 900 °C. The alkali treatment in the present study transformed the organic sulfur to readily decomposable species.

Figure 14 Changes in sulfur form distribution with residence time or distance from coal hopper for (a) parent coal and (b) alkali-treated coal of Muswellbrook.

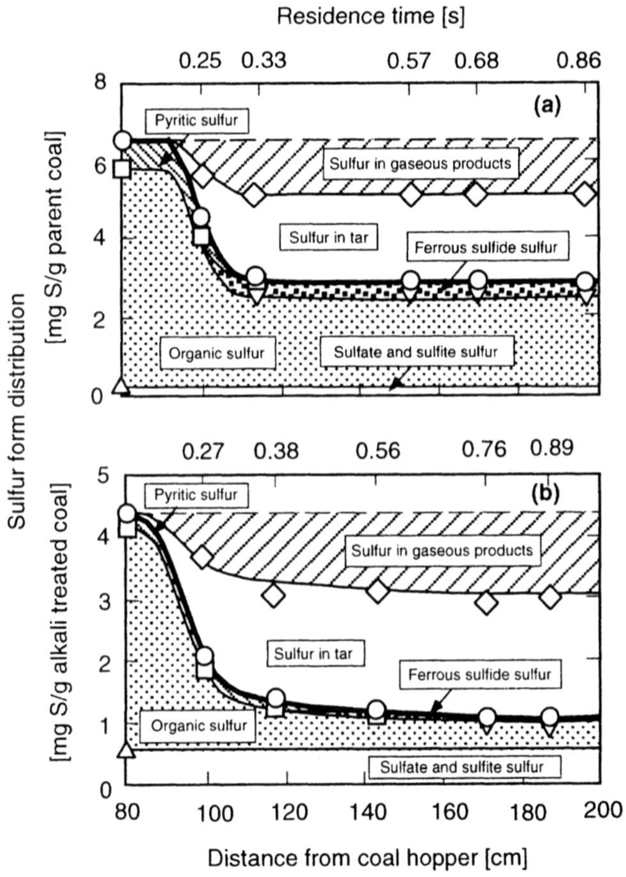

Figure 15 compares the organic sulfur contents of the parent coal, the alkali-treated sample and the char obtained by rapid pyrolysis after alkali treatment. The combined process of rapid pyrolysis with alkali treatment was effective in reducing organic sulfur contents of 0.57 and 2.2 % in Muswellbrook and Illinois No.6 parent coals to 0.04 and 0.44 % in the respective chars. For Nantong coal, however, desulfurization by the combined process was not so effective: 2.7 % of organic sulfur in the parent coal was reduced only to 2.3 % in the char. It would be necessary to permeate the coal particles with the alkali solution thoroughly and to raise the treatment temperature above 250 °C to promote conversion of the stable forms of organic sulfur in this instance.

Figure 15 Change in organic sulfur content for A, pyrolysis char; B, alkali treated coal; C, pyrolysis char with alkali tratment.

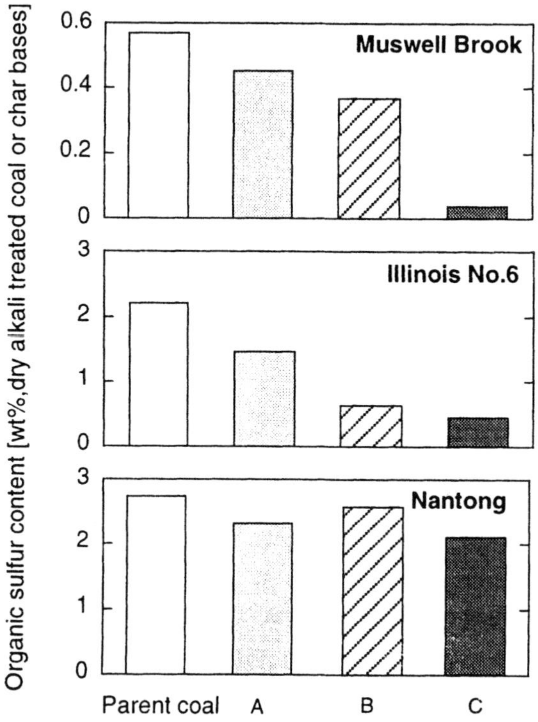

Figure 16 represents the effects of time and temperature of the alkali treatment on the organic sulfur content in alkali-treated coal and char. Although a treatment time of > 2 h (compare C and D) was not effective in reducing organic sulfur in alkali-treated coal, the organic sulfur content of the rapid pyrolysis char decreased remarkably with the increase of treatment time, probably owing to conversion of thermally stable organic sulfur to other forms which are easily decomposable in rapid pyrolysis. However, treatment at the low temperature of 200 °C for 4 h with alkali did not bring about a decrease in the organic sulfur in the pyrolysis char.

Figure 16 Effect of alkali leaching condition on organic sulfur content of Muswell Brook coal. Treatment conditions : B, 523 K, 1 h; C, 523 K, 2 h; D, 523 K, 4 h; E, 473 K, 4 h. (A is without alkali treatment. Ordinate indicates contents based on dry coal, alkali treated coal, and pyrolysis char, respectively.)

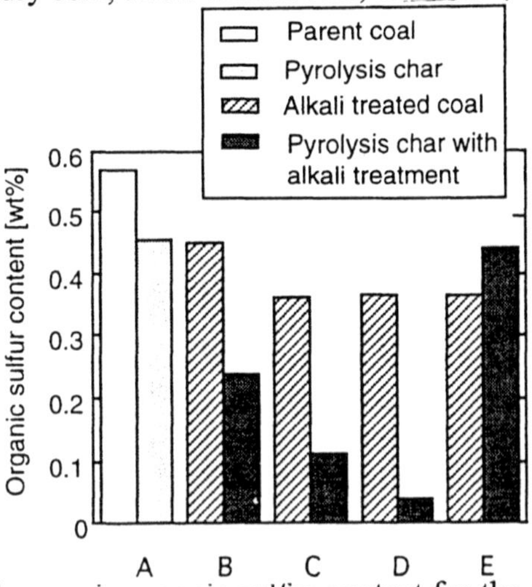

Figure 17 shows changes in organic sulfur content for the density-separated groups I and III of Muswellbrook and Nantong coals with the alkali treatment and subsequent rapid pyrolysis. The alkali treatment was carried out under the condition of 250 °C for 4 h. Though organic sulfur content of the alkali-treated group I was higher than that of group III, rapid pyrolysis reduced effectively the organic sulfur content of group I rather than that of group III for Muswellbrook. The reduction of organic sulfur proceeded for group I of Nantong coal. On the other hand, the alkali treatment was not effective for group III which showed higher sulfur content for rapid pyrolysis char than that for alkali treated sample. The amount of potassium hydroxide loaded and internal structure might be important for the transformation of thermally stable organic sulfur.

Figure 17 Change in organic sulfur content before and after rapid pyrolysis for alkali treated samples. (□, before pyrolysis ; ▒, after pyrolysis)

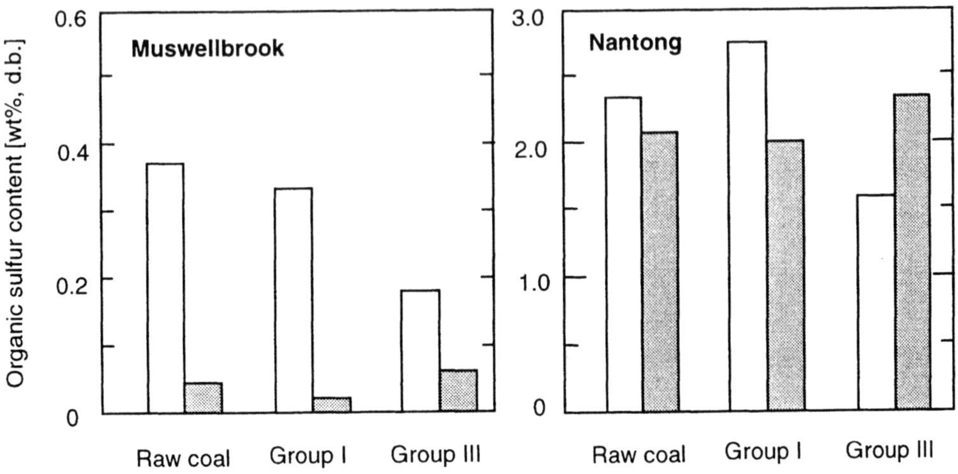

3.2. Other Chemical Pretreatments

Some chemical pretreatments were carried out as follows. The sample coals were immersed and agitated in (1) hydrogen peroxide aqueous solution, in (2) tetralin at room temperature and in (3) nitric acid aqueous solution at a boiling temperature separately, or combination of (2) and (3) [24]. After filtration and drying, rapid pyrolysis was conducted for the pretreated coals. Figure 18 shows change in sulfur content in rapid pyrolysis with various chemical pretreatments expecting oxidation of sulfur species in coal. The results of potassium hydroxide treatment are also shown in this figure for comparison. Total sulfur decreased from 10.7 % to 5.7 % during rapid pyrolysis of Mequinenza parent coal. Hydrogen peroxide treatment has not significant effect on desulfurization. On the other hand, rapid pyrolysis of nitric acid treated coal reduced the sulfur content effectively. Compared with the parent coal and the nitric acid pretreatment of Muswellbrook coal, potassium hydroxide treated sample showed greater extent of organic sulfur removal after rapid pyrolysis. However the potassium hydroxide treatment was effective for the desulfurization of subbituminous coal, this method could not apply to lignite because of dissolution. Distinct extent of desulfurization was not observed for bituminous Nantong coal in the nitric acid and potassium hydroxide treatments. The pretreatment by tetralin and nitric acid provided efficient removal of sulfur from Nantong coal though conventional methods could not eliminate organic sulfur from such a high rank coal.

Figure 18 Desulfurization with pretreatments and pyrolysis.
(, before pyrolysis ; , after pyrolysis)

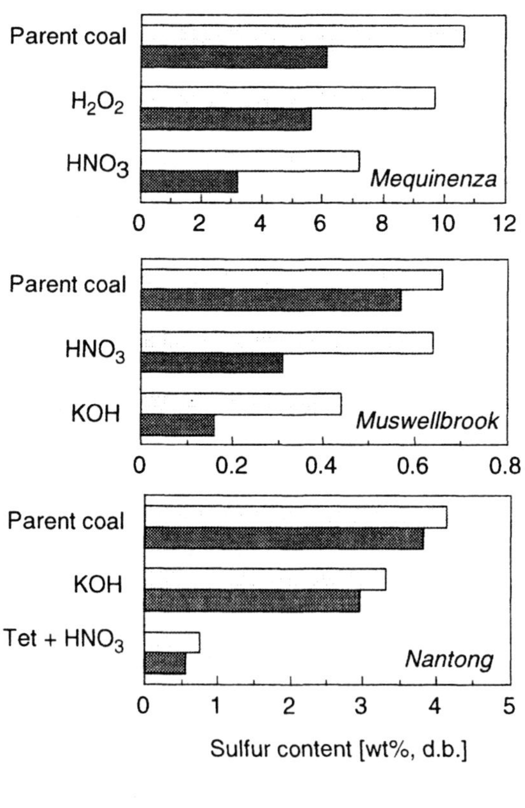

Figure 19 shows XANES spectra for the nitric acid treated Illinois No.6 coal before and after rapid pyrolysis. Strong peaks are observed around 2472 eV for the coal before pyrolysis, which are attributable to disulfide, thiophene and sulfide. A strong peak at 2474.5 eV results from sulfoxide. This peak is larger than that of parent coal. After the rapid pyrolysis of nitric acid treated coal, the peaks of disulfide and sulfide disappear and the sulfoxide peak becomes small.

Figure 19 Sulfur K-edge XANES spectra of HNO_3 treated coal before and after pyrolysis.

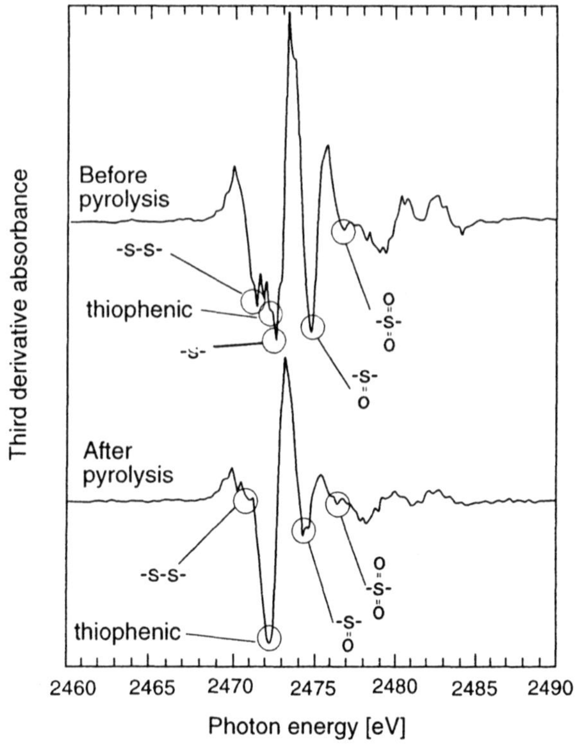

4. CONCLUSION

In a series of studies on chemical coal cleaning, the desulfurization behavior in pyrolysis was investigated for raw coals and their density separated samples. The existence of optimum desulfurization temperature in the slow heating experiments could be explained by the desulfurization scheme considering the competing phenomena between uptake reaction and dispersion of hydrogen sulfide within a particle. The extent of organic sulfur removal during rapid pyrolysis shows linear decreasing tendency with the increase of density though the extent of organic sulfur removal was different from coal to coal even with the same carbon content. A good correlation was observed between the extent of organic sulfur removal and aromatic carbon fraction for the density-separated samples. More than 90 % of organic sulfur was removed from a subbituminious coal by the combined process of rapid pyrolysis with alkali leaching.

REFERENCES

[1] Mitchell, S. C., "*Hot gas cleanup of sulphur, nitrogen, minor and trace elements*", IEA Coal Research, London, p.7 (1998)
[2] Kilbane, J. J. "*Bioprocessing and Biotreatment of Coals*" (Ed. D.L. Wise), Dekker, New York, p.487 (1990)
[3] Chriswell, C. D. Markuszewski, R. and Norton G. A., "*Processing and Utilization of High Sulfur Coals IV*", Dugan, P. R., et al., Eds., Elsevier, Amsterdam, p.399 (1991)
[4] Attia, Y. A. and Lei W., "*Processing and Utilization of High Sulfur Coals II*", Chugh, Y. P. and Caudle, R. D., Eds., Elsevier, Amsterdam, p.202 (1987)
[5] Boron, D. J. and Kollrank, R., *Min. Eng.*, **38**(2), 120 (1986)
[6] Gordon, C. R., "*Advanced Coal Cleaning Technology*", IEA Coal Research, London, p.67 (1991)
[7] Kelemen, S. R., George, G. N., Gorbaty, M. L., *Fuel*, **69**, 939 (1990)
[8] Kelemen, S. R., Gorbaty, M. L., George, G. N., Kwiatek, P. J., Sansone, M., *Fuel*, **70**, 396 (1991)
[9] Huffman, G. P., Huggins, F. E., Mitra, S., Shah, N., Pugmire, R. J., Davis, B., Lytle, F. W., Greegon, R. B., *Energy Fuels*, **3**, 200 (1989)
[10] Huffman, G. P., Mitra, S., Huggins, F. E., Shah, N., Vaidya, S., Lu, F., *Energy Fuels*, **5**, 574 (1991)
[11] Batchlor, J. D., Gorin, E. and Zielke, C. W., *Ind. Eng. Chem.*, **52**(2), 161 (1960)
[12] Robinson, L., *Fuel*, **55**, 193 (1976)
[13] Sugawara, T., Sugawara, K. and Ohashi, H., *Fuel*, **67**, 1263 (1988)
[14] Sugawara, K., Tozuka, Y., Sugawara, T. and Nishiyama, Y., *Fuel Process. Technol.*, **37**, 73 (1994)
[15] Sugawara, T., Sugawara, K., Sato, S., Chambers, A. K., Kovacik, G. and Ungarian D., *Fuel.*, **69**, 1177 (1990)
[16] Sugawara, T., Sugawara, K., Nishityama, Y., Sholes, M. A., *Fuel*, **70**, 1091 (1991)
[17] George, G. N. Gorbaty, M. L. Kelemen, S. R., Sansone, M., *Energy Fuels*, **5**, 93 (1991)
[18] Sugawara, K., Gun-ji, T. and Sugawara, T., *Energy Fuels*, **11**, 1272 (1997)
[19] Sugawara, K., Takusari, T., Sugawara, T., Shirai, M. and Nishiyama, Y., *Int. J. Soc. Mater. Eng. Resources*, 7(2), 312 (1999)
[20] Wert, C. A., Hsieh, K. C., Tsng, B. H., Ge, Y. P., *Fuel*, **66**, 914 (1987)
[21] Sugawara, K., Tozuka, Y., Kamoshita, T., Sugawara, T. and Sholes, M. A., *Fuel*, **73**, 1224 (1994)
[22] Shirai, M. and K. Sugawara, *Int. J. Soc. Mater. Eng. Resources*, 7(2), 320 (1999)

[23] Sugawara, K., Abe, K., Sugawara, T., Nishiyama, Y. and Sholes, M. A., *Fuel*, **74**, 1823 (1995)
[24] Sugawara, K., Sugawara, T., Shirai, M., *Jpn. J. Appl. Phys.*, **38**, 608 (1999)

Chapter 3

DYNAMIC THERMOPLASTIC BEHAVIOR OF COAL

Jun-ichiro Hayashi and Tadatoshi Chiba**

1. INTRODUCTION

Coal having a carbon content in the range from 82 to 89 wt% is called a caking coal, which undergoes softening and subsequent resolidification processes along with pyrolysis when heated over *ca.* 600 K. The processes are mostly governed by pyrolysis. Upon pyrolysis, coal particle converts into a three-phase mixture of gases, liquids and residual solid. Depending on composition of the mixture, the particle exhibits different softening and resolidification propensities, which in turn affects the composition and the net conversion of coal into gaseous and liquid products through the change in the extent of release of gases and liquids into surrounding liquid or gas phase. In liquefaction, coals with high softening propensity dissolve quickly into the process recycle solvent so that further hydrogenation occurs by liquid-liquid interactions, while those with little softening propensity gradually liquefy by solid-liquid interactions. In combustion and gasification, swelling [1], agglomeration, fragmentation [2], deposition onto the furnace and the inner or outer tube surfaces and reactivity [3] of char particles more or less depend on the softening propensity of parent coal, all of which affect stable operation and reactor performance. Further, in metallurgical coke making, formation and growth of the mesophase spheres depend to a large extent on the softening propensity and determine the coke properties [4].

The softening and resolidification processes have long been evaluated by the Gieseler plastmetry as a change in the fluidity with temperature and/or time. The plastometer is a constant torque plastometer [5, 6], which evaluates the fluidity of a bed of coal particles when constant torque is applied. The fluidity is expressed from measured revolution of a stirrer in the bed as the dial divisions/min (DDPM; 1 DDPM = 3.6 degrees/min) and is

* Center for Advanced Research of Energy Technology, Hokkaido University, N13, W8, Kita-ku, Sapporo, 060-8628, Japan (chiba@carbon.caret.hokudai.ac.jp).

converted into the viscosity of the bed based on calibrated relationship like DDPM/min = 10^{-7}/poise for Newtonian fluids [7, 8]. Therefore, the plastometer can detect the particle-particle interactions due to softening at least of the particle surface but not of the bulk of individual particles. Nonetheless, data from it has successfully been utilized in industrial coke production to selection of suitable coals and heating conditions for production of good-quality metallurgical coke. As the fluidity is a function of many factors, a number of empirical equations have consequently been derived, which directly correlate the fluidity of softening coal to the properties of the parent coal such as elemental and petrographic compositions, the volatile matter content, the reflectance, and so on. They have been useful enough for conventional mass production of coke featured by heating of strongly caking coals at relatively slow rate around 3 K/min. However, empirical equations are generally valid only for calculations at conditions where the data were obtained and can never be confidently applied to those at conditions other than those. Recently, such a situation actually happened in development of a new coke making process, which allows us by rapid preheating to blend non- or weakly-caking coals with caking coals. The Gieseler plastmetry is no more applicable for such cases because the plastometer hardly stands for rapid heating conditions.

Thus, efforts have been done to develop new devices such as needle penetrometer/ dilatometer [9] and *in situ* ^{1}H-NMR [10, 11], which can sensitively detect the softening and resolidification processes of coal under rapid heating. However, establishment of even a purely empirical correlation with coal properties and operating conditions on the basis of data from individual devices does require many runs of experiments. Furthermore, even if the correlation is established, it inevitably has a limitation in application, as mentioned above, and it can hardly be used for prediction of the softening and resolidification processes as a function of a wide range of parent coal properties and heating conditions. For the kind of prediction, development of mechanistic models describing the processes is necessary and thus the present article critically reviews progress and the present situation in the modeling and highlights theoretical and experimental works further needed in future.

2. Visco-Elastic Properties of Coal

Coals at softening and resolidification stages are often assumed to behave as a visco-elastic material like a solution of polymer with high molecular weight and concentration [12, 13], the visco-elastic behavior of which is well described by a Maxwell element shown in Fig.1. As is seen in the figure, the element consists of a spring having the elastic modulus, G, and a dash pot having the viscosity, η. When strains, γ_1 and γ_2, are applied respectively to the spring and the dash pot, the stress, σ, brought about by the element varies with time, t, as

$$\sigma = G\gamma_1 = \eta(d\gamma_2/dt)$$

From the above equation, the rate of change in the strain, γ, of the element is written by

$$d\gamma/dt = d\gamma_1/dt + d\gamma_2/dt = (1/G)(d\sigma/dt) + \sigma/\eta$$

Figure 1 Maxwell model for visco-elasticity.

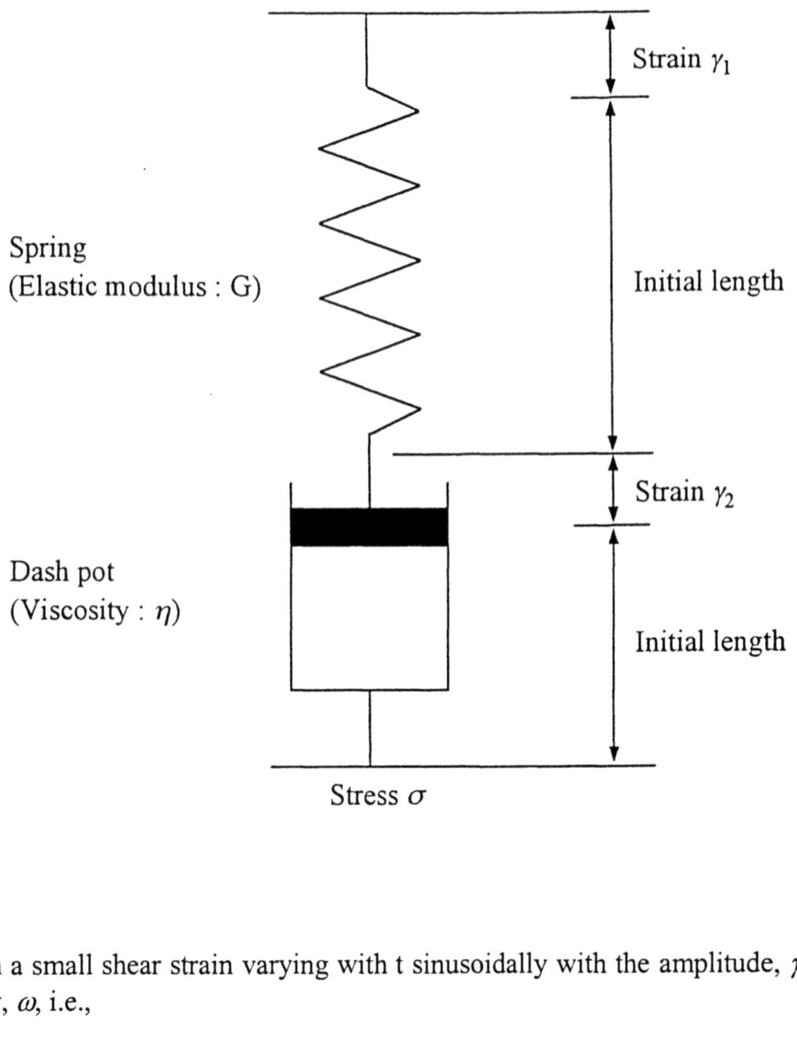

When a small shear strain varying with t sinusoidally with the amplitude, γ_0, and the frequency, ω, i.e.,

$$\gamma = \gamma_0 e^{i\omega t}$$

is applied to the element,

$$d\sigma/dt + (G/\eta)\sigma = Gi\omega\gamma_0 e^{i\omega t}$$

The general solution of the above equation is

$$\sigma = \sigma_E + i\sigma_V$$

$$\sigma_E = \{\omega^2\tau^2/(1+\omega^2\tau^2)\}G\gamma_0\{\cos\omega t + C\exp(-t/\tau)\}$$

$$\sigma_V = \{\omega\tau/(1+\omega^2\tau^2)\}G\gamma_0 \sin\omega t$$

where τ is the relaxation time defined by η/G and the integral constant, C, is determined with the initial condition that both σ_E and σ_V are independent of an arbitrary integer, n, at $t=2n\pi/\omega$ as

$$C = 0$$

and hence, the dynamic strain is given by

$$\sigma = G\{i\omega\tau/(1+i\omega\tau)\}\gamma$$

Based on the above equation, the complex modulus, G_C defined as σ/γ is expressed by

$$G_C = G\{i\omega\tau/(1+i\omega\tau)\} = G_E + iG_V$$

Here, G_E and G_V are the elastic and viscous moduli, respectively, written by

$$G_E = G_C\{\omega^2\tau^2/(1+\omega^2\tau^2)\} = |G_C|\cos\delta$$

$$G_V = G_C\{\omega\tau/(1+\omega^2\tau^2)\} = |G_C|\sin\delta$$

with the lag of the strain behind the stress by an angle of δ defined by

$$\tan\delta = G_V/G_E$$

Recently, Nomura *et al.* [14] applied the above theory to analyze the visco-elastic properties of coals, using a rheometer equipped with a mechanically oscillating device briefly shown in Fig.2. Putting a pellet of 20 mm in diameter and 2.1 mm in thickness of coal powder between the two parallel discs of 25 mm in diameter, they applied a continuous sinusoidally varying strain with an amplitude of 0.1 % and detected the strain upon heating at 3 K/min up to a holding temperature of 723 K to obtain G_E, G_V and tan δ as a function of temperature and frequency. Though having difficulty to detect proper strains at temperatures below the softening temperature determined by dilatometry, they found that the observed dependencies of the above visco-elastic properties of coals on temperature and frequency are qualitatively in accordance with those of polymer solutions. Along with softening, G_E and G_V decrease while tan δ increases. Both G_E and G_V exhibit minima of 10^4 to 10^5 Pa while tan δ attains a maximum near the temperature at which the maximum fluidity in the Gieseler plastmetry is obtained. The maximum of tan δ exceeds unity for caking and weakly caking coals but tan δ falls below unity as the temperature rises over the resolidification temperature determined by the Gieseler plastmetry. They further suggested that the relative compressibility of semicoke can be evaluated based on the difference in G_E among coals near the resolidification temperature.

Figure 2 Device for detection of dynamic thermoplastic properties of coal tablet.

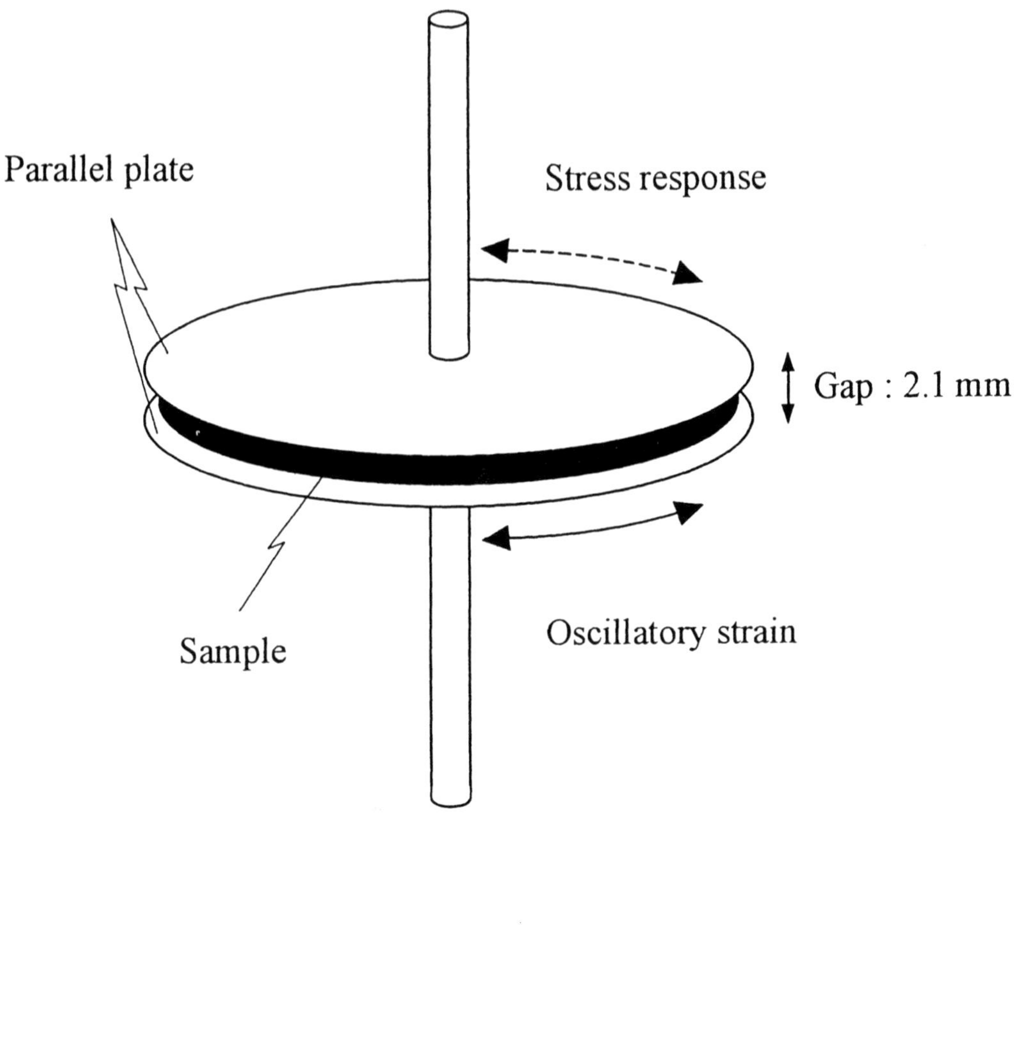

In spite of the efforts using advanced instrumentation, the results obtained seem to be in principle comparative to those from the Gieseler plastmetry. They could explain most of their data in correspondence with the characteristic temperatures defined in the Gieseler plastmetry. The changes of the visco-elastic properties, e.g., G_E, G_V and tan δ, surely reflect what happens in the softening and resolidification processes and G_E and its change with temperature may give a new insight particularly to change with temperature and dependency of semicoke structure on parent coal nature upon heating. As mentioned in the introduction, softening and swelling coal is a three-phase mixture of gases, liquids and solid char. However, it is at present not clear whether changes of the visco-elastic properties with temperature can quantitatively be described in terms of the structure and composition of three-phase mixture due to very limited knowledge particularly of the structural and compositional change with time upon heating as well as the instantaneous properties of the liquid and solid involved in the mixture.

3. Modeling of Softening Coal

It is a widely accepted fact that softening and subsequent resolidification of caking coal is accompanied by pyrolysis and carbonization reactions. Though coal has a very complicated chemical structure and a number of reactions could take place upon heating, a simplified reaction model was first proposed by van Krevelen and coworkers [15, 16]. They assumed the softening due to formation of metaplast from coal, which is followed by resolidification due to conversion of metaplast to semicoke. The consecutive reaction proceeds accompanying gas evolution and is expressed as

$$
\begin{array}{c}
\text{gas} \\
k_1 \quad \uparrow k_2 \\
\text{coal} \rightarrow \text{metaplast} \rightarrow \text{semicoke}
\end{array}
$$

where k_1 and k_2 represent the reaction rate constants for the metaplast formation and decomposition, respectively. Fitting the kinetic model to experimental data of gas evolution from coal upon heating at constant rate or at constant temperature after heating, k_1 and k_2 were optimized for four kinds of coals and several kinds of polymers. From the results of analysis, they concluded that a good quality of coke can be produced if the activation energies for k_1 and k_2 are approximately equal. They also suggested that the metaplast must have a wide molecular weight distribution containing a sufficient amount of hydrogen or a portion of low molecular weight components and behave as a liquid at temperatures at least equal to or higher than the softening temperature. Though implied that fractions extractable by certain solvents from softening coal constitute a portion of the metaplast, they never definitely indicated how to identify and quantify the metaplast.

Solomon *et al.* [17] proposed a systematic method to predict the fraction of the metaplast or the liquid fraction in softening coal. They employed polymer concepts to calculate the molecular weight distribution during pyrolysis of coal. Describing coal by macromolecular network statistics, they predicted the distribution by their functional group (FG) model for gas evolution combined with their network decomposition model for the elemental processes of depolymerization vaporization and cross-linking (DVC) and determined the liquid fraction from the distribution. Then, without any data to compare, predicted results of the liquid fraction were utilized to estimate the viscosity of softening coal, assuming that coal is an inhomogeneous reacting polymer melt, i.e., slurry. They adopted Mooney's equation [18, 19] for the slurry viscosity, η, as a typical two-phase model, which is written with the viscosity of the liquid, η_l, the Einstein coefficient, k_E, the volume fraction of solids in slurry, ϕ_s, the critical volume fraction of solids at which η goes to infinity, as

$$\ln(\eta/\eta_l) = k_E\phi_s/(1-\phi_s/\phi_c) \tag{1}$$

They further adopted the Andrade equation for η_l

$$\eta_l = \eta_{l0}\exp\{E_V/(RT)\} \tag{2}$$

which leads Eq. (1) to

$$\eta = \eta_{l0}\exp\{E_V/(RT)\}\exp\{k_E\phi_s/(1-\phi_s/\phi_c)\} \tag{3}$$

Assuming fixed values of the model parameters in Eq. (3) together with two adjustable parameters for description of coal structure empirically on the basis of literature data, the fluidity predicted was compared with data obtained with a Gieseler plastometer. From fairly good agreement between them, they concluded that the model for softening coal structure and the slurry equation (Eq. (3)) are applicable to prediction of the liquid fraction and the coal fluidity, respectively.

The above conclusion stimulates further work to justify the assumptions in their modeling. The comparison made between predicted and observed results seems to be indirect because, as mentioned above, the Gieseler plastmetry gives the viscosity indirectly through conversion of DDPM to the viscosity by experimental calibration, which is usually necessary for each coal and measurement condition to examine the Weissenberg effects involved in the plastmetry [20]. Therefore, directly measured viscosity data are vital to examine the assumptions made in the prediction and to refine the model. It is also needed to further examine the applicability of the slurry equation (Eq. (2)). For example, the Einstein coefficient, k_E, and the critical volume fraction, ϕ_c, assumed in the prediction were 5.0 and 0.65, respectively, which were arrived at empirically for the predicted fluidity to fit the data. k_E theoretically derived for slurry consisting of liquid and monodispersed rigid spheres at low concentration [21] is 2.5, which is a half of that assumed. The use of 5.0 for k_E was justified by assuming the situation where liquid is entrapped within large agglomerates of solid. On the other hand, the value of ϕ_c seems to be reasonable because $\phi_c = 0.64$ for the random close packing of monodispersed sphere system and ϕ_c can range up to 0.9 for close packing of polydispersed particle systems [22]. However, examination of physical validity of the values of k_E and ϕ_c should be done after ϕ_s is properly given because difference in both assumed parameters from the theoretical values for ideal conditions might be resulted in if the liquid fraction, $1-\phi_s$, is assumed to constitute only a portion of the metaplast as done for solvent extractable fractions.

4. Measurement of Coal Viscosity and Quantification of Metaplast

As mentioned above, it is necessary for examination of the softening model developed by Solomon et al. to measure directly and systematically the viscosity of softening coal. The authors [9] developed a needle penetrometry, which had been applied to measurement of the viscosity of polymer melts and polymer solution [23]. Figure 3 illustrates a schematic diagram of the penetrometer/ dilatometer (P/D meter). It equips a linear variable differential transformer as a position sensor for a 2.0 mm ϕ needle and a 10 mm ϕ piston to measure the penetration depth and the volumetric swelling of coal, respectively. Samples for the measurements were prepared by pelletizing pulverized coal particles with a high mechanical pressure loading. The pellet had a diameter of 10 mm and a height raging from 3 to 8 mm with voidages as low as less than 0.05. Nearly the same characteristics of the needle penetration and the dilation were observed as those for samples prepared by lathing from coal lump as a block having the same size as the pellet. The vertical positions of the tip of the needle, H, and the volumetric dilation ratio of the pellet, Q, were determined as

$$H = H_0 - H_t$$

$$Q = H_p / H_0$$

from measured values of the initial pellet height, H_0, the vertical distance of the tip of the needle from the pellet bottom at a time t, H_t and the height of pellet, H_p. Further, in order to apply an equation of motion for a needle descending through a Newtonian polymer melt under gravity containing no foam or bubble [9];

$$h(dh/dt) = \{mg/(2\pi\eta)\}\ln(a/b) \tag{4}$$

where m, a and b denote the mass of needle, the diameter of needle and the diameter of pellet, respectively, the net penetration depth was defined by

$$h = (QH_0 - H_t)/Q \tag{5}$$

Figure 3 Schematic diagram of needle penetrometer/dilatometer.

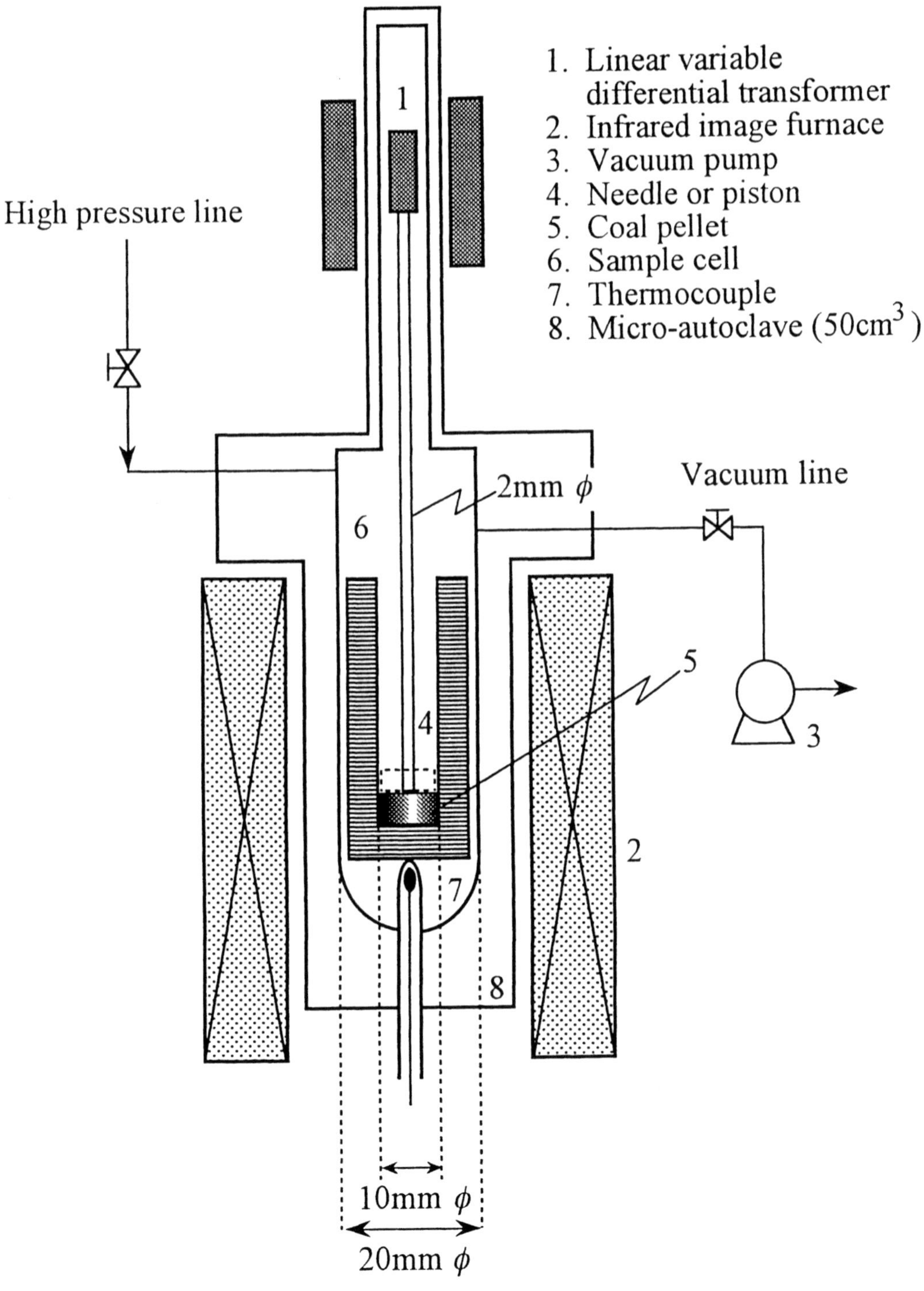

In Eq. (4), the softening pellet is assumed to behave a Newtonian fluid so as to reduce the number of the parameters to be optimized for an experimental data. This assumption for simplicity will later be discussed.

Figure 4 shows typical changes of H, Q and h for 5.2 mm high Goonyella coal pellet heated at different rates up to 773 K and held at the temperature, T_h, for a while. From the changes, it is clear that P/D meter can detect sensitively the effect of the heating rate on softening characteristics of the coal pellet; at heating rates of 2, 3 and 10 K/min, H and h fall off at about 5 mm, suggesting termination of the needle penetration near the bottom of the pellet, while, at the slowest heating rate of 1 K/min, the termination occurs at h = *ca.* 3 mm before the needle reaches the pellet bottom. Similar sensitivity was attained for different coals listed in Table 1, as shown in Fig.5 where h is plotted against temperature, T. It should be noted that an appreciable penetration does occur at the higher heating rate for BYC and CER coals, both of which exhibit little penetration propensity at the slower heating rate. Detection of the effects of T_h and gas pressure, P, is also possible by P/D meter, as shown in Fig.6. It is shown that, though the needle penetration commences at the same temperature, deeper and more rapid penetration takes place at higher T_h and that more intensive penetration occurs at higher P for coals having a high volatile content possibly due to confinement of a larger portion of pyrolysis products within the pellet at higher P. Based on the equation of motion, Eq. (4), changes with T of the viscosity of coal, η, were calculated by analyzing the net penetration curves. Figure 7 represents the effect of the heating rate on η calculated from the curves shown in Fig.4. It is seen that η decreases with T rapidly in a range from 10^{12} to 10^4 Pa·s and the decrease lasts up to higher temperatures for higher heating rates. At each heating rate, a minimum is obtained in η *vs.* T. However, those except for that at 1 K/min are apparently obtained. At 1 K/min, the penetration terminates at a height 2 mm above the bottom with deceleration of needle descent due to resolidification (Fig.4). On the other hand, at the higher heating rates, the deceleration and the termination of penetration take place near the bottom of the pellet as the needle tip reaches the bottom. The latter deceleration at h =*ca.* –5 mm disappeared for thicker pellets. Though not shown in Fig.7, the minimum of η decreased with the heating rate [24]. Different changes of η with T were also obtained for different coals as shown in Fig.8. The onset of penetration at lower temperature and the larger extent of decrease in η sensitively reflect more intensive caking propensities of coal.

Table 1 Relevant properties of typical coal samples so far used.

		Ultimate Analysis					Proximate Analysis		Gieseler Fluidity Parameters				Pellet	
Coal	(Abbr.)	C	H	O[a]	S	N	Ash	Volatile	ST[d]	NFT[e]	FT[f]	log(MF/ddpm)	H_o	ρ_o
		[d.a.f.[b] wt%]					[m.f.[c] wt%]		[K]	[K]	[K]	[--]	[mm]	[g/cm³]
Akabira	(AKA)	80.7	5.9	10.5	0.6	2.3	4.2	40.6	620	705	733	2.50	5.6	1.14
Cerrejon	(CER)	82.3	5.5	10.1	0.5	1.6	1.0	37.7	n.d.	n.d.	n.d.	n.d.	6.1	1.19
Black Water	(BLW)	86.8	4.9	5.8	0.4	2.1	7.7	23.4	687	725	752	2.51	5.4	1.20
Goonyella	(GOO)	87.4	4.9	5.2	0.6	2.0	7.3	26.4	681	735	769	3.25	5.2	1.05
Byron Creek	(BYC)	86.0	4.7	7.5	0.3	1.5	11.2	23.8	700	725	746	0.48	5.3	1.23
Peak Downs	(PDH)	88.4	4.9	3.9	0.6	2.1	10.0	21.4	695	743	775	2.56	4.9	1.30

a; by difference, b; dry-ash-free basis, c; moisture-free basis, d; softening point temperature, e; temperature given for maximum fluidity, f; solidification point temperature

Figure 4 Changes with temperature of vertical position of needle tip, H, volumetric dilation ratio, Q, and net needle penetration depth, h, for AKA coal pellet heated at different rates up to 773 K and 1.0 MPa-N_2.

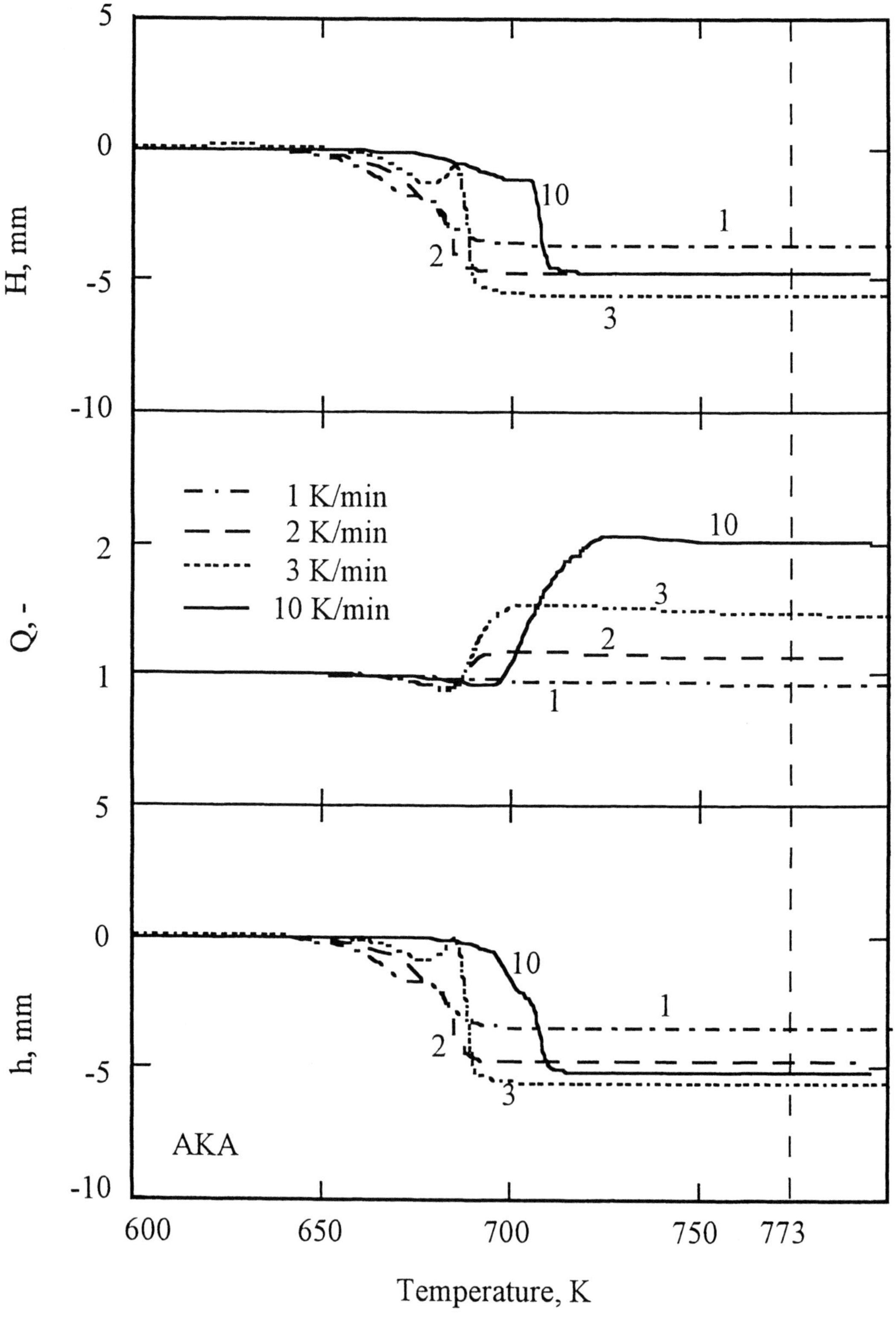

Figure 5 Changes with temperature of net needle penetration depth, h, for pellet of different coals heated at (a) 3 K/min and (b) 10 K/min up to 773 K under 1.0 MPa N_2 gas atmosphere.

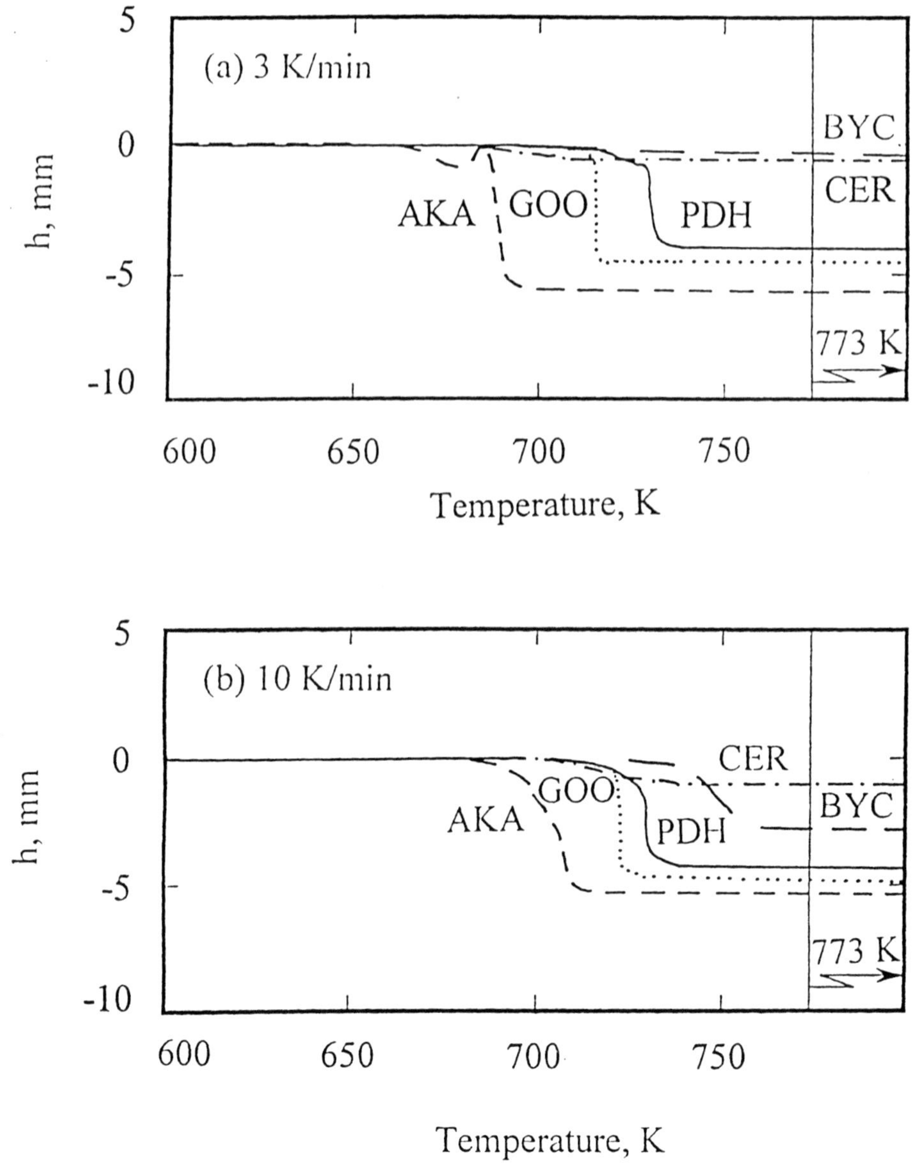

Figure 6 Effects of (a) holding temperature and (b) pressure on net needle penetration depth, h, for different coal pellets heated at 10 K/min.

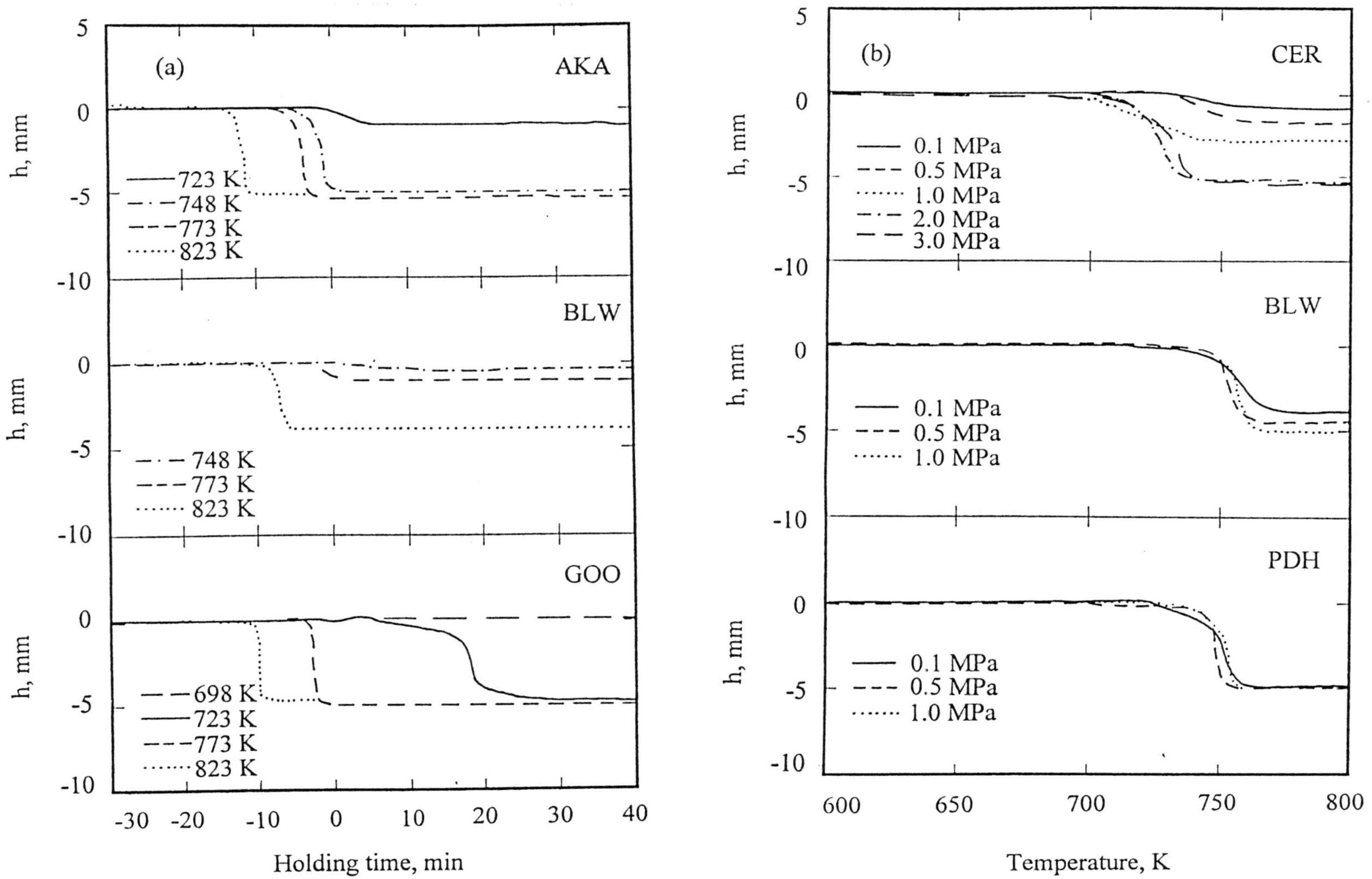

Figure 7 Variation of apparent viscosity, η, with temperature for AKA coal pellet heated at different rates under 1.0 MPa N_2 gas atmosphere.

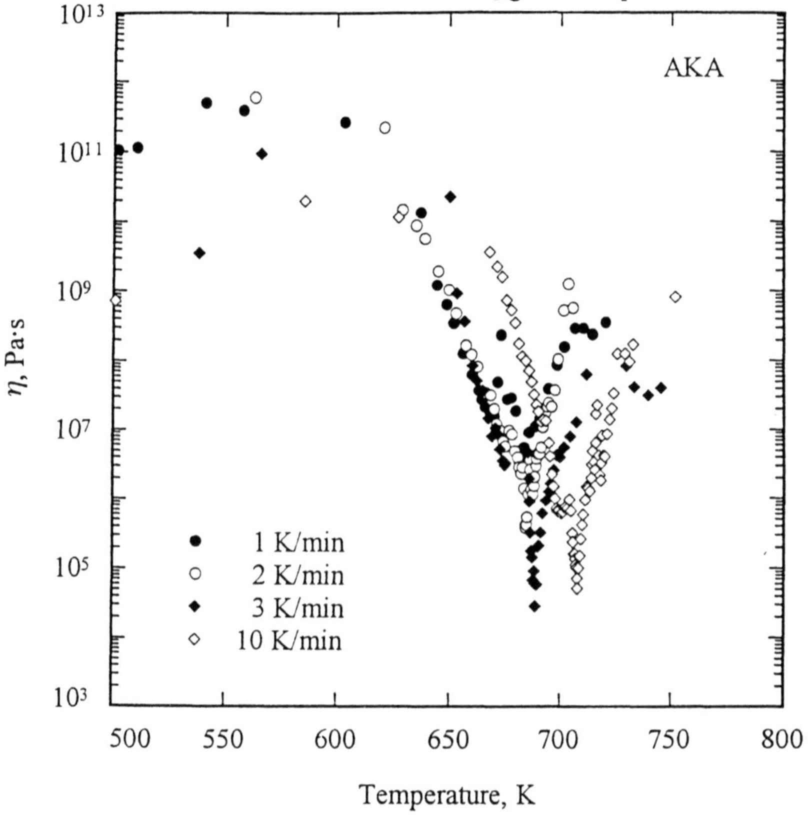

Figure 8 Variation of apparent viscosity, η, with temperature for various coal pellets heated at 10 K/min under 1.0 MPa N_2 gas atmosphere.

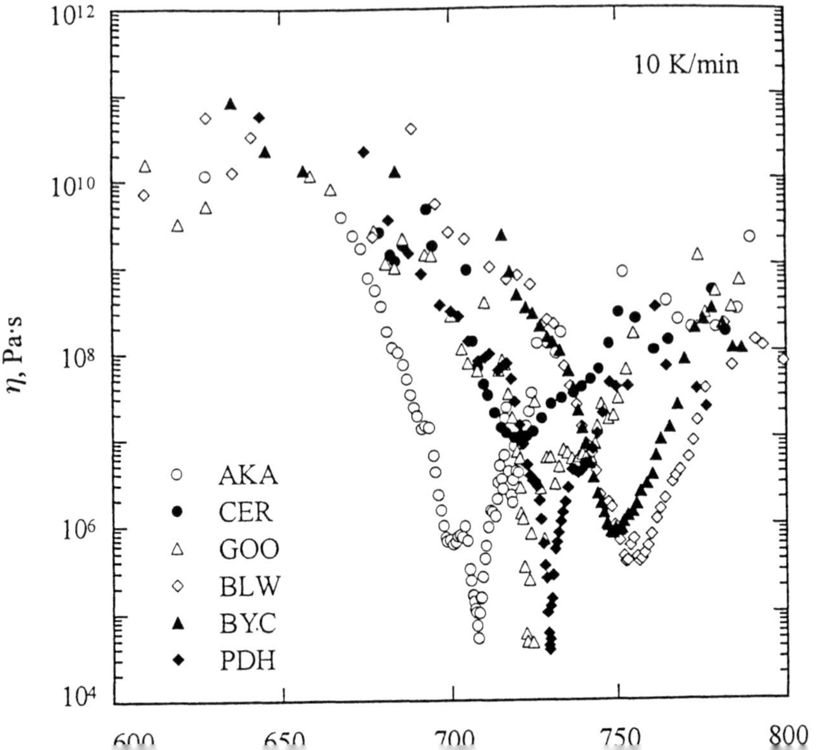

Figure 9 Schematic diagram of apparatus for preparation of pellets subjected to solvent extraction.

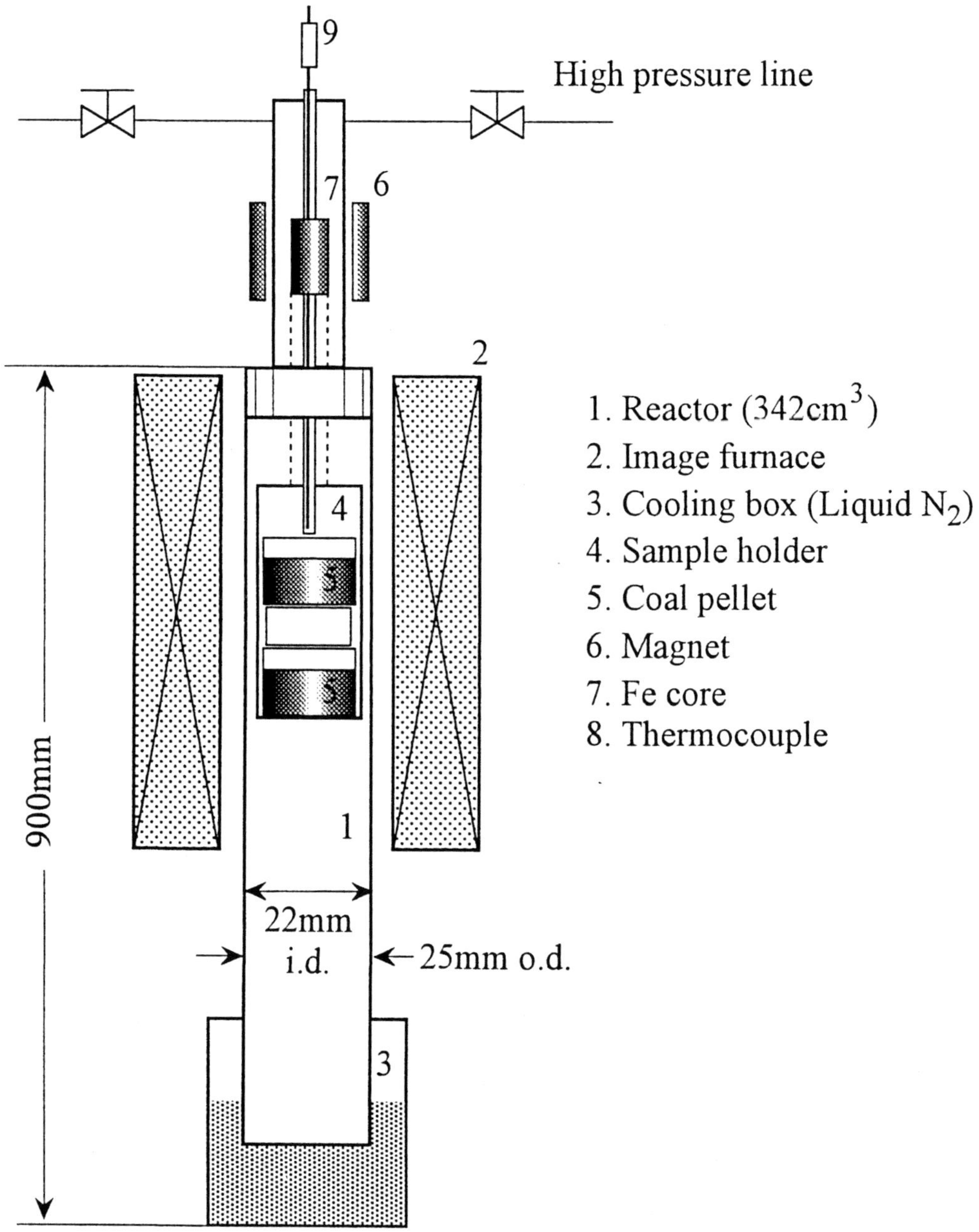

The above described changes of η with T corresponded to those in the fraction of the metaplast candidates contained in the pellet such as solvent extracts, components giving endotherm in a differential scanning calorimetry (DSC) and mobile hydrogen analyzed by ^{1}H-NMR technique. A schematic illustration is shown in Fig.9, which was designed to measure the yield of solvent extracts in the pellet upon heating. A pair of coal pellets used for the needle penetrometry was fixed in the holder magnetically hanged in a tubular reactor and heated in the same manner as in the needle penetrometry. When the pellet temperature rose to a prescribed one, the pellet was dropped to the reactor bottom dipped in a bath of liquid nitrogen to minimize progress of pyrolysis and carbonization upon cooling down to room temperature. The pellets were then subjected to extraction by solvents with different extractabilities and the extract yields were determined by weighing the extract residue as mass fractions of the mass of the recovered pellet. Figure 10 shows the yield of pyridine extract, Y_P, as a function of temperature, T, for Goonyella coal pellets heated at different rates and for different coals heated at 10 K/min. It is seen that change of Y_P appreciably depends on the heating rate as well as coal nature. A maximum in Y_P appears at each heating rate (Fig.10(a)). The maximum yield increases and the temperature at which the yield is attained rises with the heating rate. On the other hand, the maximum yield attained for different coals at the same heating rate (Fig.10(b)) appears at higher temperature for coal having a lower minimum viscosity as shown in Fig.5. Correspondence was thus examined between change of Y_P and η with T. A typical result is shown in Fig.11. It is clear that both changes correspond well with each other but the temperature for the minimum viscosity apparently disagrees with that for the maximum yield.

In the DSC measurement, the pellet was loaded in an aluminum cell and heated at 10 K/min up to 823 K in a 80 ml/min nitrogen gas flow to sweep volatile matter released by pyrolysis. The heat-treated pellet was once cooled down to room temperature and heated again at the same condition as above. Figure 12 compares change with T in the heat flux upon the first heating for Akabira coal pellet with that of the viscosity. There can be seen two endothermic peaks at temperatures around 350 K and 680 K. The former peak could be attributed to release of water remaining in the pellet while the latter to the phase transition along with conversion of coal into metaplast. Since no peak was detected in the repeated heating, revealing that the change of the heat flux with T is irreversible. The DSC technique seems to be useful to detect the temperature related to the maximum yield of metaplast but to be impossible to quantify the fraction of metaplast in the pellet, even if the observed endotherm could be arisen only from the formation of liquid.

Figure 10 Variation of pyridine extract yield, Y_P, with temperature for (a) GOO coal pellets heated at different rates and (b) for different coal pellets heated at 10 K/min under 1.0 MPa N_2 gas atmosphere.

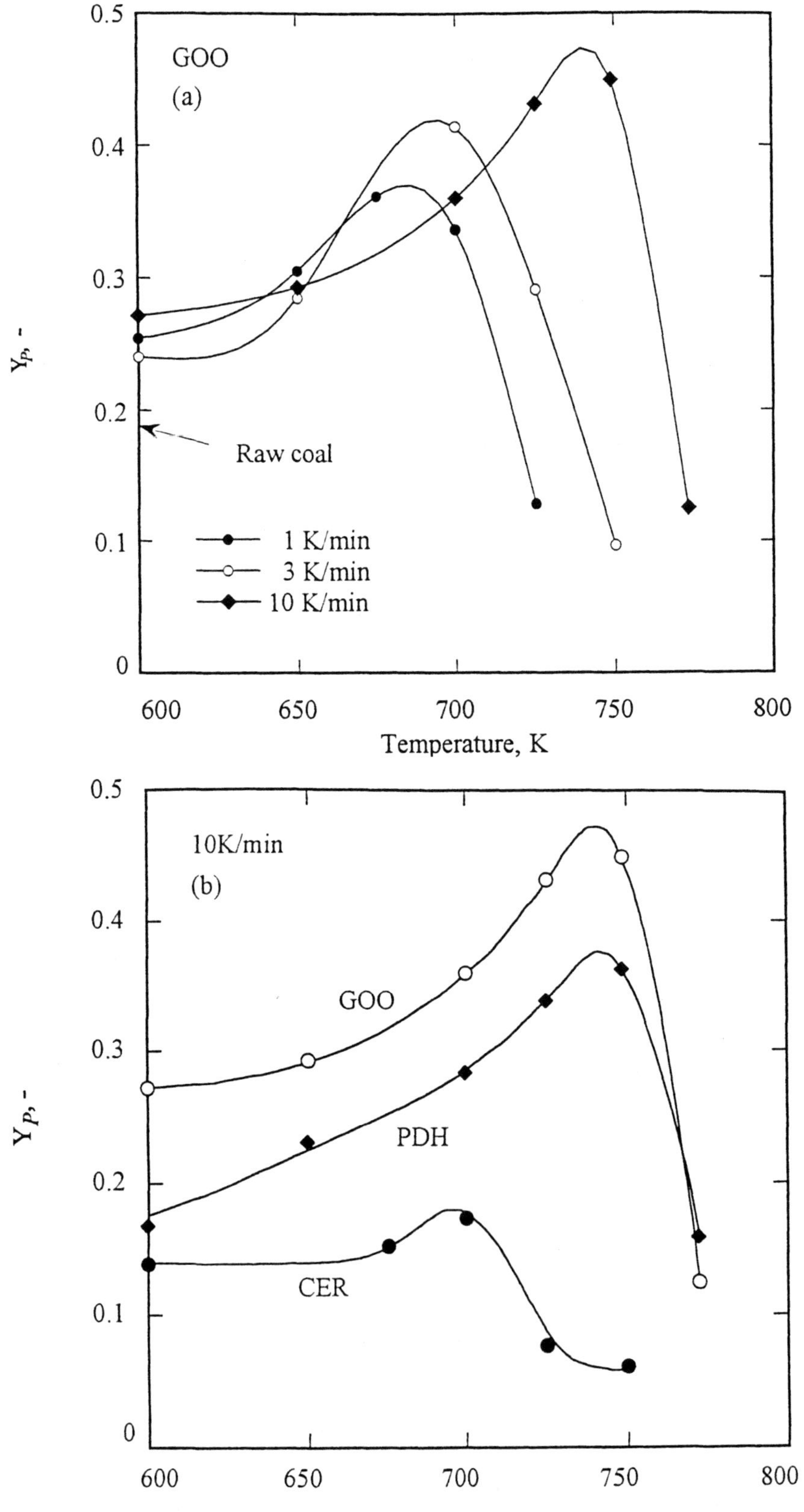

Figure 11 Comparison between change with temperature of apparent viscosity, η, and that of pyridine extract yield, Y_P, for GOO coal pellet heated at 10 K/min under 1.0 MPa N_2 gas atmosphere.

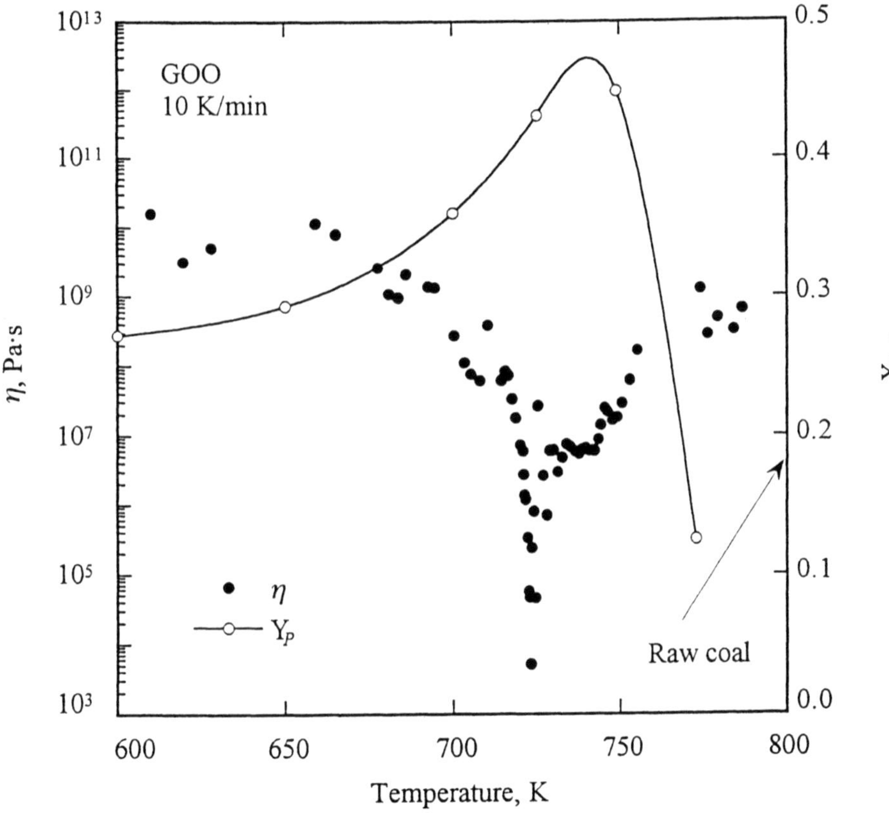

As suggested by Chermin and van Krevelen [16], the metaplast would be rich of hydrogen, which could be detected by ^{1}H-NMR technique [10, 11]. Thus, spin-spin relaxation signals were also obtained by *in-situ* ^{1}H-NMR measurements for the coal particles with a Bruker model (CXP-40 Pulse Spectrometer) [24]. A solid-echo method was employed at a resonance frequency of 38.27 MHz with a 90 deg pulse width of 5 μs and a delayed-time for a pulse interval of 15 μs to quantify change in the atomic fraction of mobile hydrogen in the pellet with time at the same conditions as above. The spin-spin relaxation signals could be deconvoluted into those arisen from two components showing a slow exponential decay and a rapid Gaussian decay. A typical result is shown in Fig.13. In general, relaxation of mobile hydrogen existing in liquid is expressed by an exponential decay function with a longer spin-spin relaxation time, T_2, while that of immobile or rigid hydrogen in solid is done by a Gaussian decay function with a shorter T_2. Here, the atomic fraction of mobile hydrogen, f_{Hm}, is calculated by

$$f_{Hm} = I_L(0)/\{I_L(0)+I_G(0)\}$$

where $I_L(0)$ and $I_G(0)$ represent the initial signal intensities arisen from the mobile hydrogen and the immobile hydrogen, respectively. Figure 14 plots f_{Hm} against T observed for Akabira coal heated at 4 K/min and compares the change with that of the viscosity, η, of the pellet heated at 10 K/min. In the figure, f_{Hm} is seen to increase with T above 550 K, which is followed by a decrease in the higher temperature range giving a maximum of 0.40. The temperature range for the increase is lower than that for the decrease of η but the temperature for the maximum is very close to the minimum of η despite of the different heating rates due to operating limitation of the NMR apparatus. However, the approximate coincidence of the temperatures is apparent as shown in the next section.

Figure 12 Comparison between change with temperature of apparent viscosity, η, and that of DSC heat flux for GOO coal pellet heated at 10 K/min under 1.0 MPa N_2 gas atmosphere.

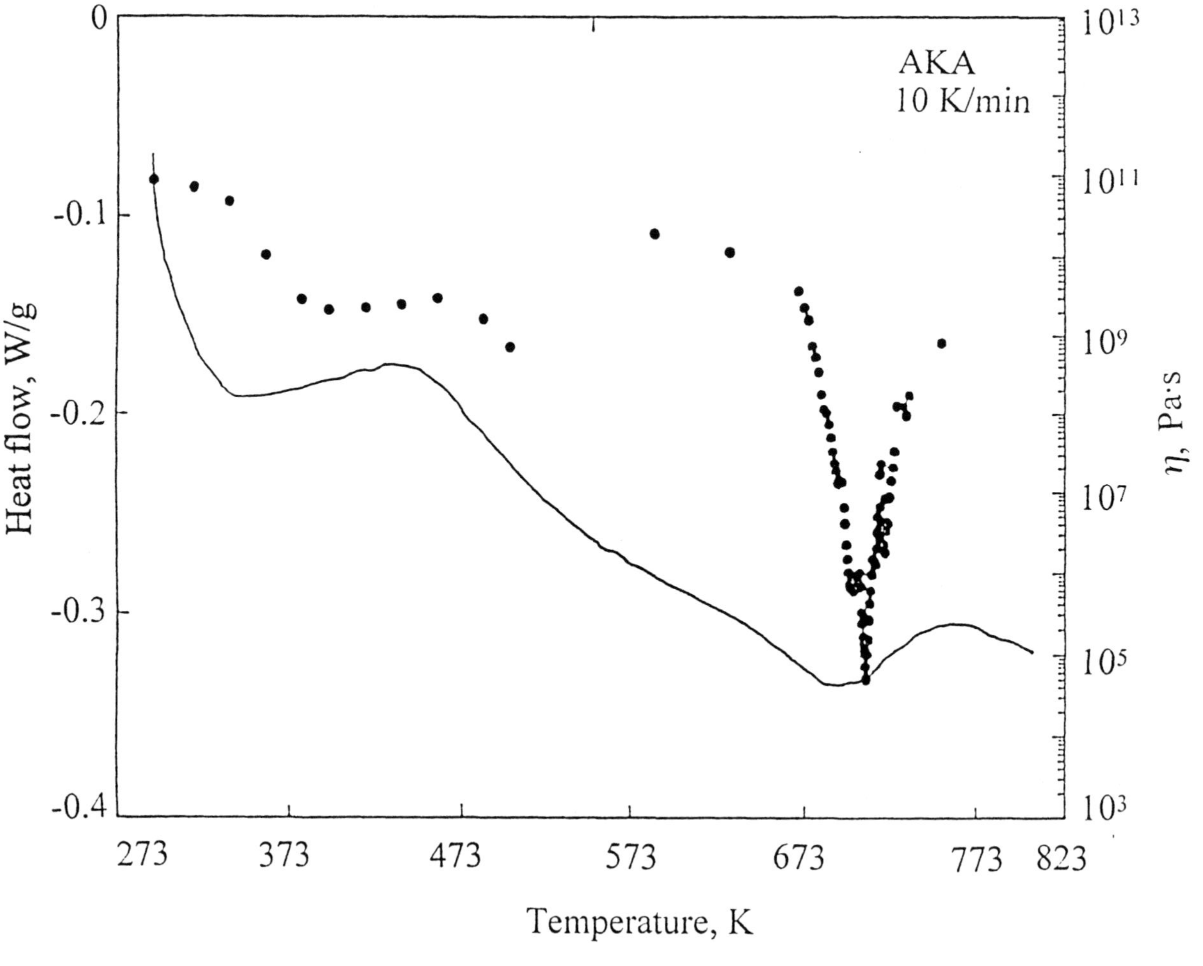

Figure 13 Typical relaxation signal observed and deconvolution of signal into those arisen from two components.

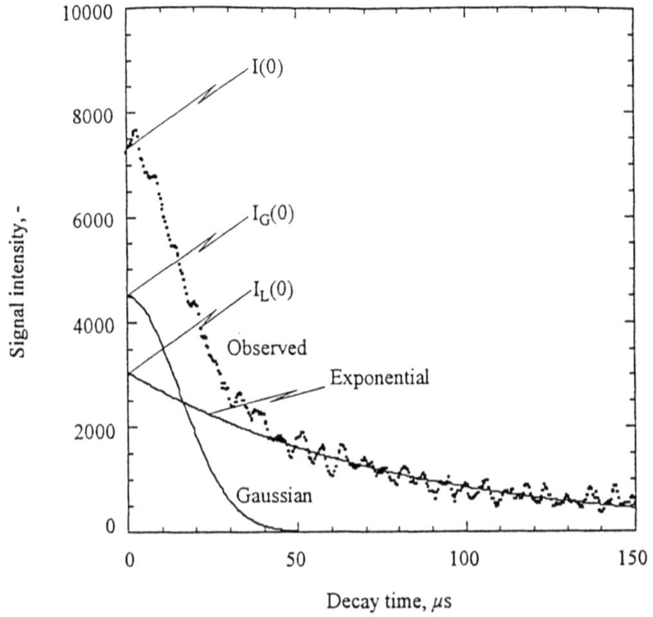

Figure 14 Comparison between change with temperature of apparent viscosity, η, and that of mobile hydrogen fraction, f_{Hm}, for GOO coal pellet heated at 10 K/min under 1.0 MPa N_2 gas atmosphere.

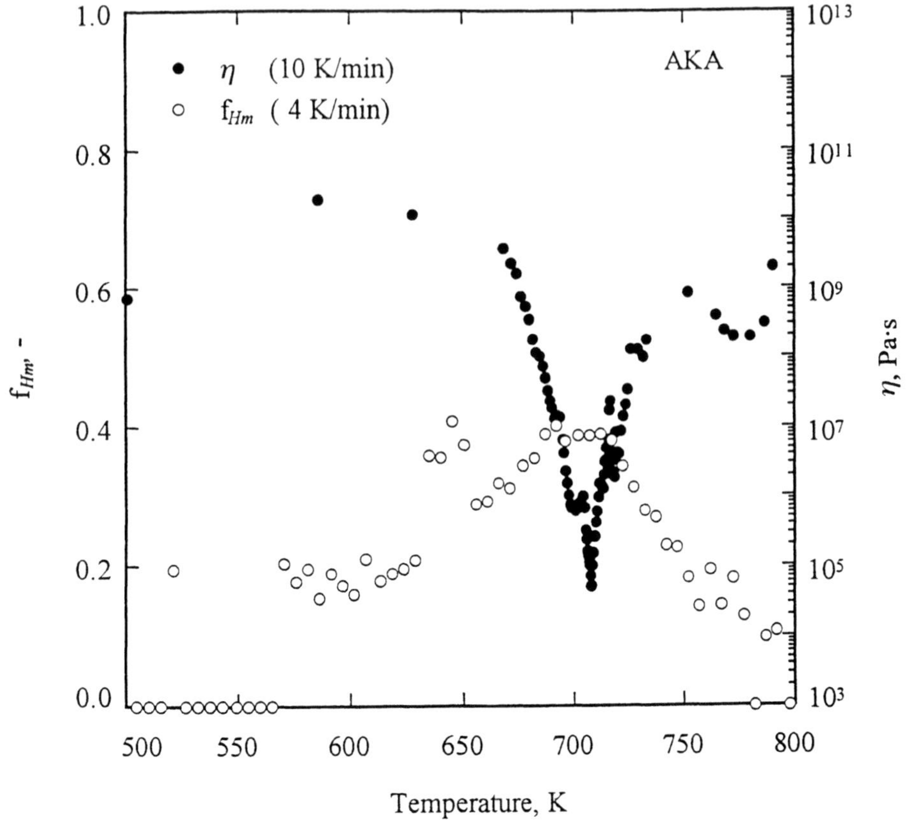

5. Experimental Examination of Slurry Model

Apart from definition of the metaplast, the apparent viscosity of slurry, η, is generally expressed as a function of the liquid viscosity, η_l, and the volume fraction of solid particles, ϕ_s, as

$$\eta = \eta_l f(\phi_s) \tag{6}$$

A number of the functions have so far been proposed for $f(\phi_s)$ and Mooney's equation [18, 19], Eq.(1) was employed by Solomon *et al.* [17] and the authors [25]. According to the above equation,

$$d\eta/dT = f(\phi_s)(d\eta_l/dT) + \eta_l\{df(\phi_s)/d\phi_s\}(d\phi_s/dT) \tag{7}$$

This suggests that the temperature at which a minimum viscosity is obtained is not necessarily equivalent to that at which a maximum fraction of liquid in the slurry is attained upon heating. Thus, verification of the validity of slurry model, *i.e.*, Eq. (6) in general or Eq. (1) was rather difficult on the basis of all the above described results from unsteady-state temperature experiments and was left for further examination. On the other hand, when a slurry is held at a temperature, η depends only on time, t, through the change in ϕ_s with t while leaving the liquid viscosity, η_l, virtually invariable without significant variation in the liquid molecular composition. In such a case,

$$d\eta/dt = 0 \text{ when } d\phi_s = 0 \tag{8}$$

and the time for the minimum of η must coincide with that for the minimum of ϕ_s. Hence, in our recent study [26], applicability of a slurry model, Eq. (1) was first examined in a series of experiments employing model coals that consist of semicoke mixed with different amounts of pyridine extract and then Eq. (8) is examined experimentally to quantify the metaplast.

Figure 15 illustrates changes with time of the height of the tip of needle, H, volumetric ratio of the pellet at a time t to the initial pellet, Q, and net needle penetration depth, h, for GOO coal pellets heated to the different holding temperatures, T_h. It is seen that the needle penetration takes place at all T_h. At T_h = 700 and 710 K, the penetration is accompanied by the pellet dilation while, at 690 K, Q remains virtually constant after a slight shrinkage of the pellet due to the initial volatile release. For all T_h, the net penetration, *i.e.*, the change of h, commences after the pellet is heated to the temperature. The apparent viscosity of the pellet, η, was calculated from the changes in h with t, as described. The result is shown in Fig.16. With heating, η is seen to decrease rapidly from 10^9 to 10^6 Pa·s. At higher T_h the rate of decrease is larger and hence the decrease occurs earlier. A minimum of η appears at each appears at each T_h. The minimum viscosities, η_{min} are attained at 690 and 700 K when the needle tip passes through nearly a half way down to the pellet bottom while at 710 K near the bottom of the pellet. The appearance of the latter minimum is considered as a result of local resolidification of the pellet bottom because a further decrease of η was observed for pellets with larger initial height. Hence, analyses of the viscosity data hereafter are confined to those obtained at 690 K, where the minimum of η can be obtained during the temperature holding.

Figure 15 Changes in vertical position of needle tip, H, volumetric dilation ratio, Q, and net needle penetration depth, h, for GOO coal pellets with holding time, t, at different holding temperatures, T_h.

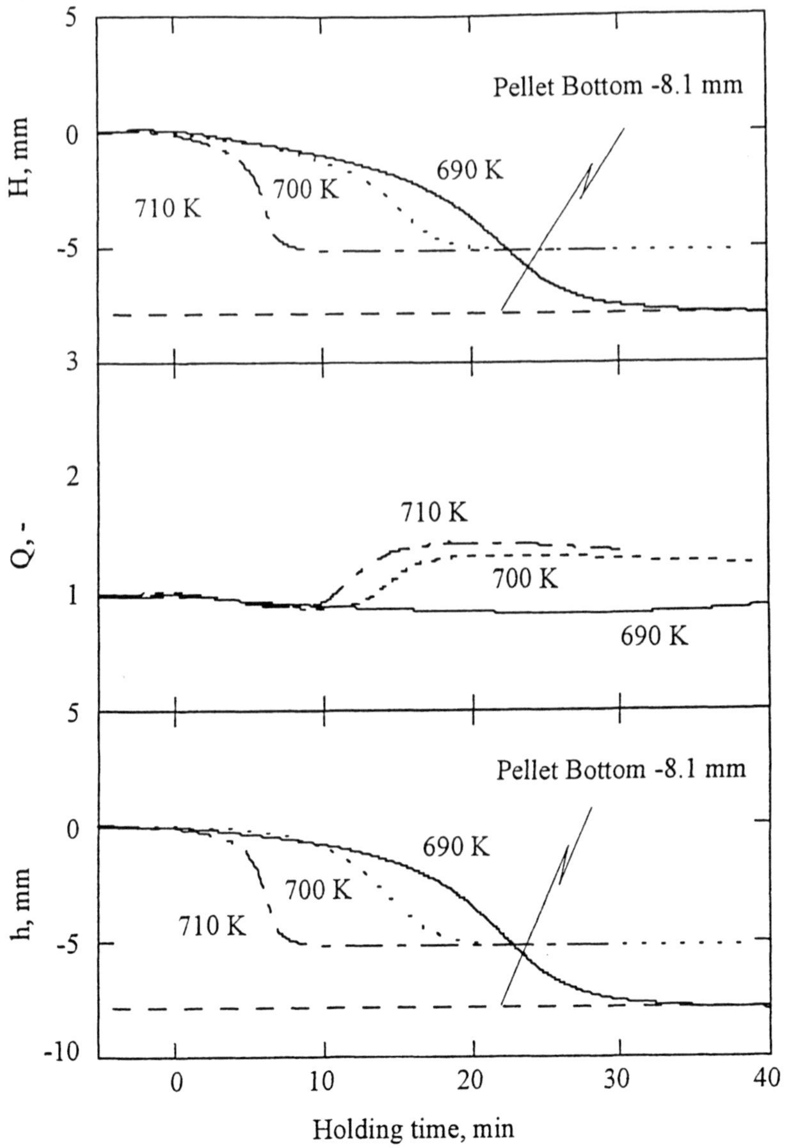

Figure 16 Changes in η for GOO coal pellets with holding time, t, at different holding temperatures, T_h.

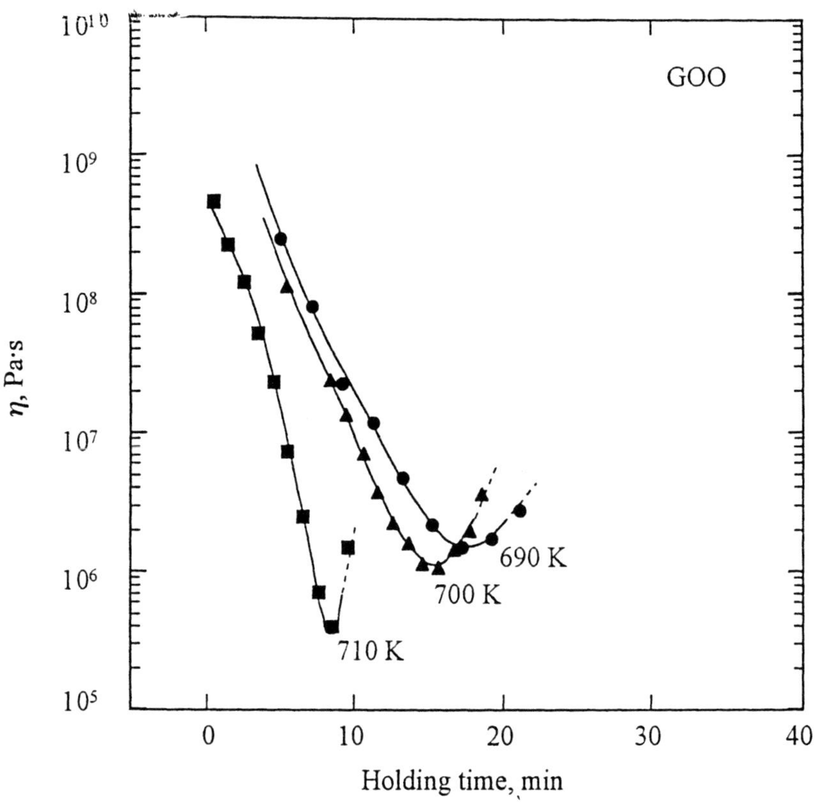

Figure 17 shows the net penetration depth, h, as a function of temperature, T, for the pellets consisting of semicoke/pyridine extract mixtures with known initial mass ratios, ϕ_{Po}. It should be noted that penetration takes place for pellets with $\phi_{Po} > 0.3$ and that the rate of penetration increases with ϕ_{Po}, suggesting that pyridine extract behaves as a liquid at temperatures above 550 K. The penetration curves in the figure were analyzed using Eq. (4) to obtain the apparent viscosity, η, as a function of temperature, T, assuming a pseudo-steady state of the pellet in each narrow temperature range. Figure 18 demonstrates the result where $\ln\eta$ is plotted against 1/T from expectation that pyridine extract behaves as a liquid and its viscosity variation with T follows the Andrade equation, Eq. (2). As expected, nearly parallel linear relationships are obtained between them and the values of k_V and E_V averaged for the straight lines are $1.50{\cdot}10^{-27}$ Pa·s and 340 kJ/mol, respectively. The mass fraction of the pyridine extract in the pellet upon heating was different from that at the initial, *i.e.*, ϕ_{Po}, due to mass release from the extract. Then, the extract (ϕ_{Po} = 1.0) was heated at 10 K/min and nitrogen pressure of 1.0 MPa, while the mass release from it was measured by a thermogravimetry. In Fig.19 is shown the mass of the extract normalized by its initial mass, W/W_o, as a function of T. Since the mass release from the semicoke was negligible upon the heating up to 690 K, the mass fraction of the extract at a given temperature, referred to as ϕ_P, is calculated as follows

$$\phi_P = \phi_{Po}(W/W_0)/[1 - \phi_{Po}\{1 - (W/W_0)\}]$$

When $f(\phi_s)$ is independent of temperature, Equation (7) is simplified as

$$d\eta/dT = f(\phi_s)(d\eta_l/dT) \quad (9)$$

which implies that the temperature dependency of η is the same as that of the liquid viscosity. The data shown in Fig.18 were further examined on the basis of Mooney's equation, Eq. (1), assuming that a softening pellet consists of pyridine extract as a liquid and semicoke as a solid and pyridine extract has the same density as the residual solid, i.e., the volume fraction of solid, ϕ_s, being given as

$$\phi_s = 1 - \phi_P$$

Figure 17 Changes in h for pyridine-extract/semicoke pellets with temperature at heating rate of 10 K/min.

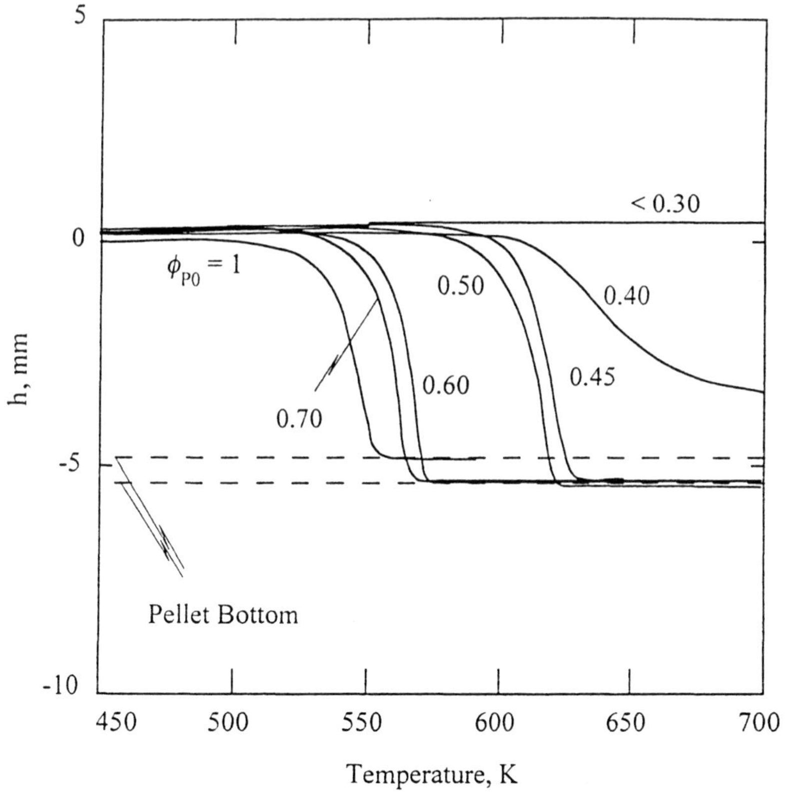

Figure 18 The Andrade plots for η_l of pyridine-extract/semicoke pellets with different initial mass fraction of extract, ϕ_{Po}.

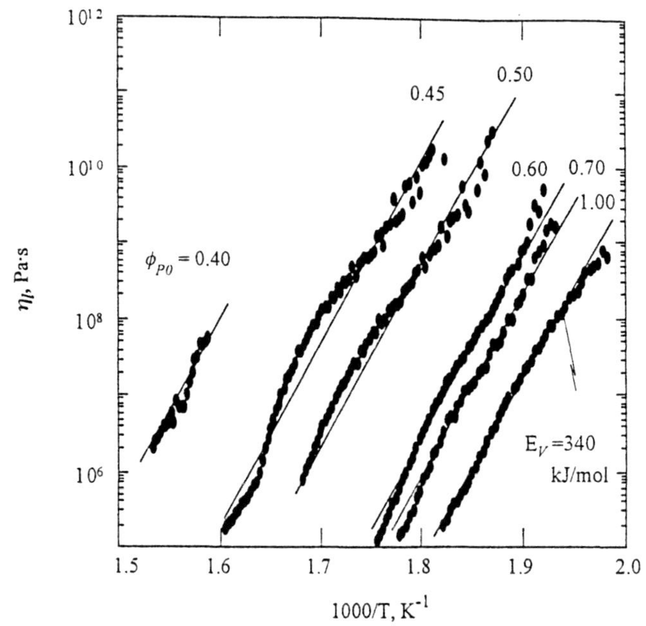

Figure 19 Change in mass fraction of pyridine-extract with temperature, T, and holding time, t.

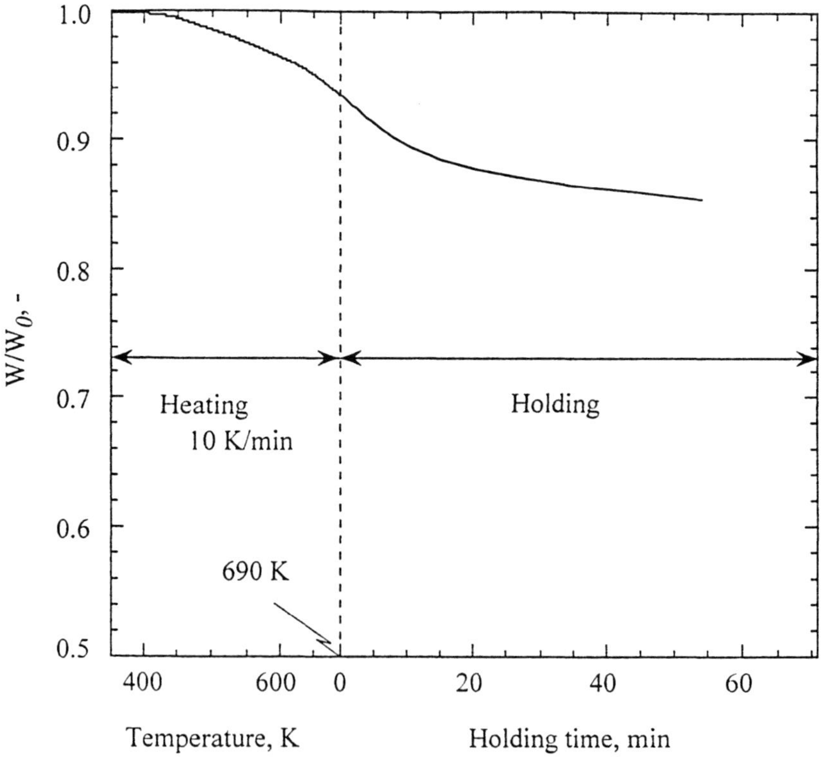

Arisen from Eq. (1), $1/\{\ln(\eta/\eta_l)\}$ is plotted against $1/\phi_s$ in Fig.20. Here, the values of η were fixed as 10^7, 10^8 and 10^9 Pa·s. The corresponding values of ϕ_s were graphically determined from the lines drawn in Fig.19 while those of η_l at the corresponding temperatures were calculated by Eq. (2) with the values of k_V and E_V for pyridine extract. As seen in the figure, the plot gives a reasonably well linear relationship. This reveals that Eq. (1) is applicable for description of softening of the model pellets. The Einstein coefficient, k_E, and the critical solid volume fraction, ϕ_c, in the equation thus determined are 5.9 and 0.77, respectively [24].

Figure 20 Relationship between $1/\{\ln(\eta/\eta_l)\}$ and $1/\phi_s$ for pyridine-extract/semicoke pellets.

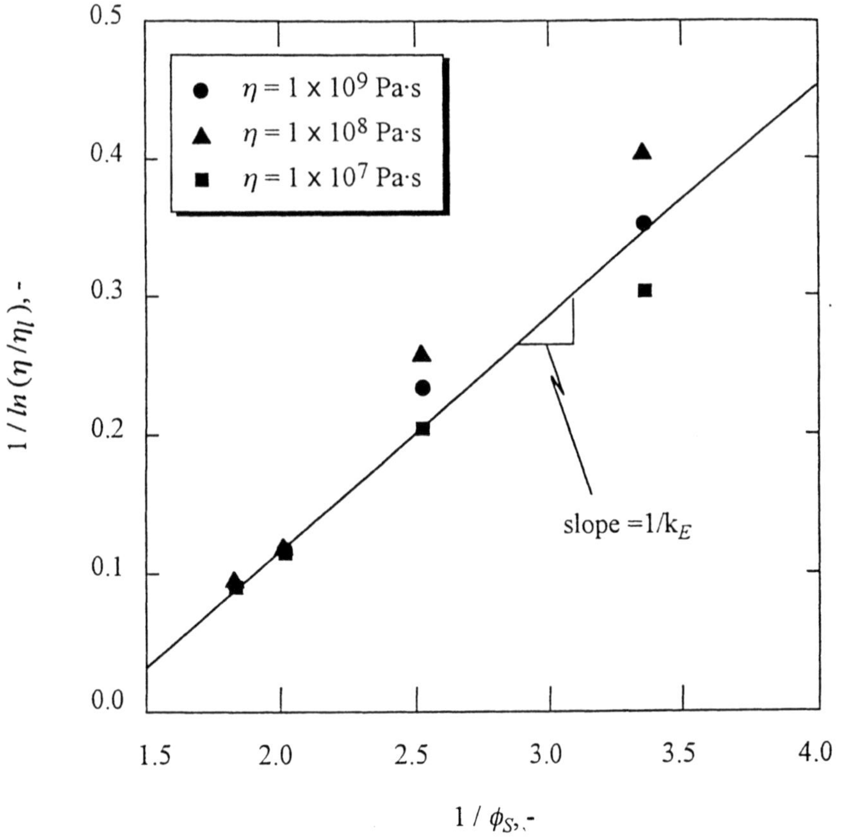

On the basis of the slurry model (Eq. (1)), correspondence of time-dependent change in with that in ϕ_s was experimentally examined. Figure 21 compares the change of η with that of the yield of pyridine extract, Y_P, for GOO coal pellet at T_h = 690 K. It is clearly demonstrated that Y_P decreases monotonously with time and never attains a maximum when the viscosity reaches a minimum. This indicates that Eq. (8) does not hold when pyridine extract is assumed to represent metaplast. As also shown in the figure, the same conclusion is derived even when quinoline or 1/1 volume ratio mixture of pyridine and CS_2 is employed as a solvent being more effective than pyridine for extraction. In addition, all the yields are significantly lower than the liquid mass fraction expected from the slurry model, $Y_{l,calculated}$, with the parameters determined for the pellets of semicoke and pyridine extract, which is also drawn as a broken line in the figure. There may be at least two possible explanations for the disagreement. First, changes would occur in chemical composition of pyridine extract with time. Second, the extract would contribute to softening as a constituent of whole metaplast. To examine the first explanation, measurements were conducted on the elemental composition by a conventional method and the molecular mass distribution by a laser-ionization mass spectroscopy for extracts heated for different period at a holding temperature. The results are shown in Fig.22. In the figure, no significant changes can be detected and the disagreement can hardly be attributed to the change in chemical composition of the extract upon heating. The above discussion indicates that the definition of the metaplast as the solvent extracts is not reasonable so far as considering that the softening coal behaves as slurry. The second explanation was examined by applying an *in-situ* ^{1}H-NMR. As mentioned above, the analysis can provide the fraction of liquid in the softening coal being heated as the atomic fraction of mobile hydrogen existing in a liquid phase that shows exponential transverse magnetization decay distinct from Gaussian decay of rigid hydrogen in a solid phase. According to previous investigations on application of ^{1}H-NMR thermal analysis to coal characterization, change in Gieseler fluidity can qualitatively be correlated with that in the molecular mobility of coals upon heating in terms of the half-width of spectrum [27, 28], the empirical second moment of spectrum [29] and the fraction of mobile hydrogen [30]. In Fig.23 change in the mobile hydrogen fraction, f_{Hm}, with time, t, is compared with $Y_{l,calculated}$ shown in Fig.21. It should be noted that the change in f_{Hm} with t has a maximum at approximately the same t as η reaches the minimum. It is also seen that f_{Hm} is slightly higher than $Y_{l,calculated}$ and this would be due that f_{Hm} is the hydrogen fraction while $Y_{l,calculated}$ is the volume fraction. The result reveals that the change of metaplast may be described well by that of mobile hydrogen though a slight difference is still left between the values of f_{Hm} and $Y_{l.calculated}$.

Figure 21 Comparison of yields of pyridine-, quinoline- or pyridine-CS_2 mixture extracts at 690 K with η and $Y_{l,calculated}$ for GOO coal pellet.

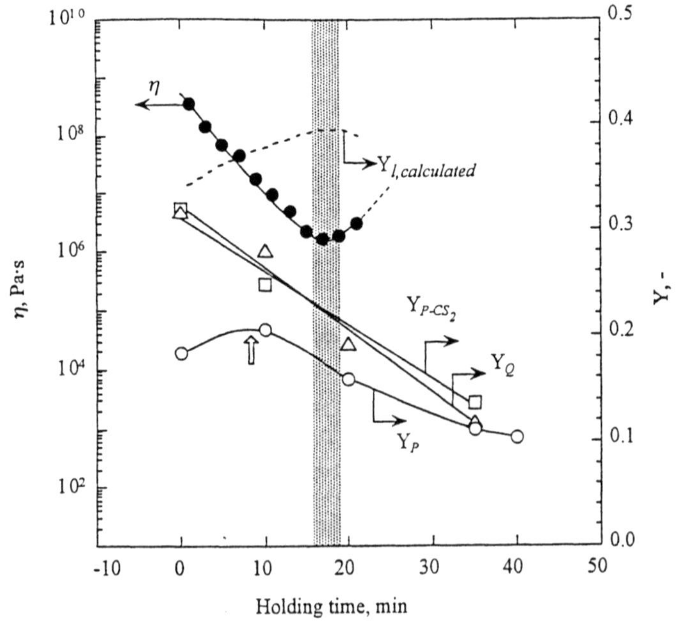

Figure 22 Elemental contents of pyridine-extract as a function of holding time at 690 K and mass spectra of pyridine-extract before and after heating at 690 K for GOO coal pellet.

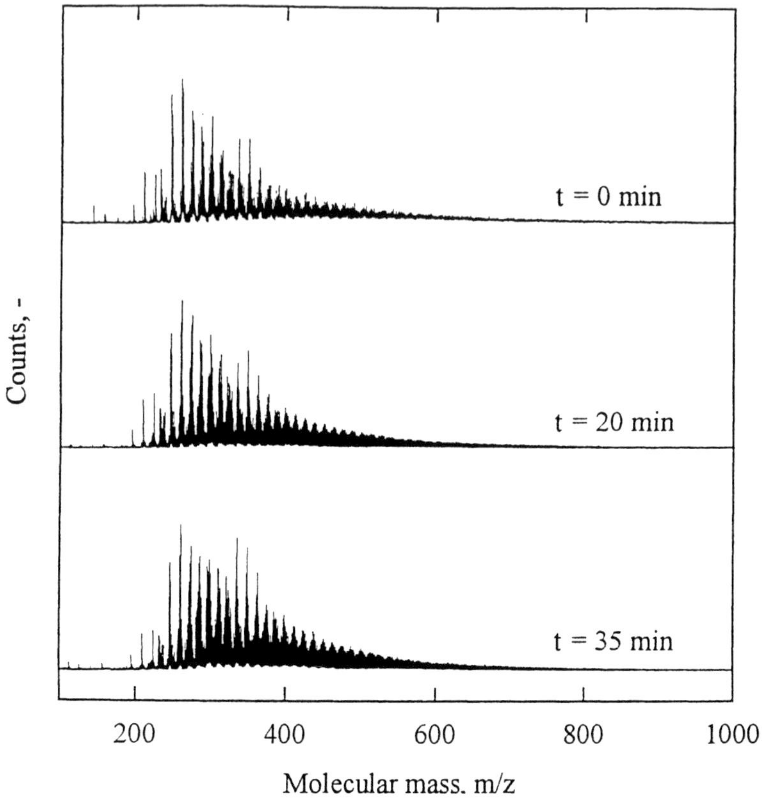

Figure 23 Comparison of f_{Hm} with $Y_{l,calculated}$ at 690 K for GOO coal pellet.

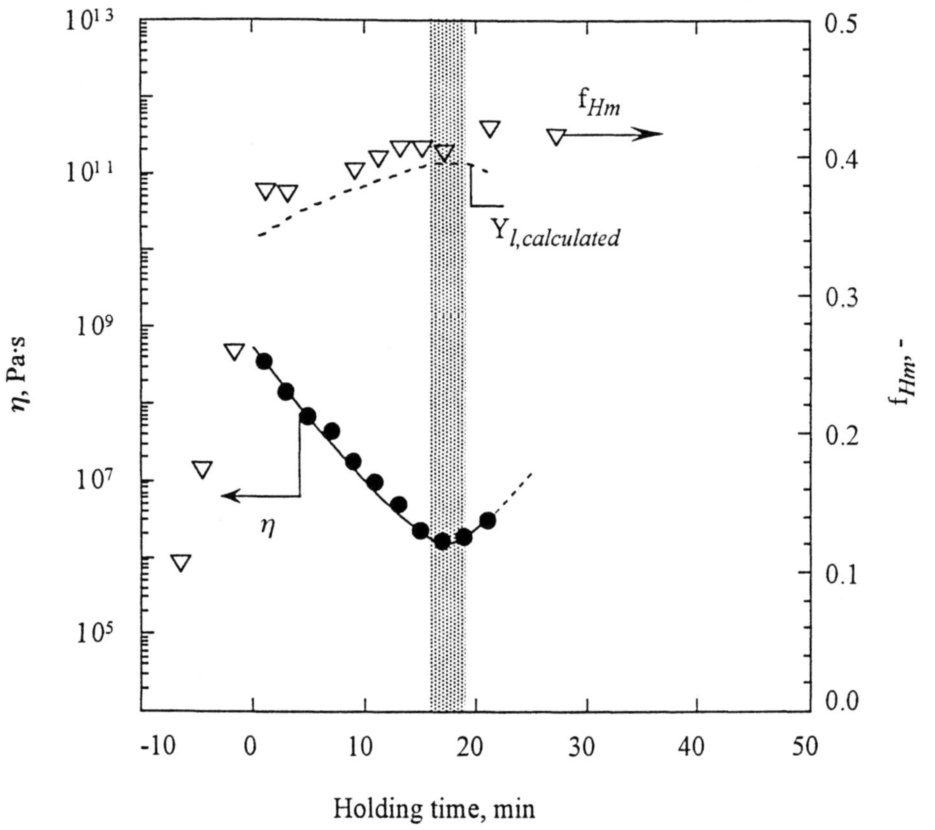

6. Concluding Remarks and Future Work

The Gieseler plastometer has widely been employed in metallurgical coke making industries, giving useful information for selection of raw coals having properties suitable for production of coke with the desired structure. In plastmetry, the softening propensity of coal is determined by measuring the rate of rotation of a stirrer with constant torque in the bed of coal particles. This means that it does not directly give us the viscosity of the coal itself. The rate of rotation must be translated to the viscosity by calibration using liquids with known viscosity and further the viscosity of the bed of coal particles must be converted to that of the coal itself. Therefore, knowledge from the plastmetry will remain useful but just empirical. The recent development of a dynamic mechanical oscillation device for visco-elasticity measurement, on the other hand, might give us more scientific and generalized coal properties in an analogy to polymer and polymer solution. Unfortunately, at present, the data reported seem to suffer from the effects of dilation of the coal tablet and slip of the oscillating disc from the tablet surface. In addition, it is not clear how to model the observed change in the visco-elastic properties with operating variables and parent coal nature.

Measurements of needle penetration into a coal pellet and volumetric dilation of the pellet with a needle penetrometer/dilatometer have enabled us to investigate details of softening and resolidification rate processes of coal upon heating. The apparent viscosity of the bulk of coal, η, can be defined from the net needle penetration characteristics on the basis of the equation of motion for the needle. The present paper further examines applicability of an existing general slurry model and quantification of the metaplast, which has long been thought responsible for the processes, in quantitative relation to the slurry model. As a result, the following are concluded within the experimental findings so far obtained.

1. The slurry model here employed (Eq. (1)) is basically applicable to evaluation of η for pyridine-extract/semicoke mixtures upon heating.
2. The portions of coal extracted by solvents such as pyridine, quinoline and mixture of pyridine and CS_2 behave as liquid. However, their yields of the extracts are appreciably lower than that expected from the model, $Y_{l,calculated}$ as that of the metaplast and the change in η with time at a steady temperature does not correspond to those in their yields.
3. The change of η much better corresponds with that of the atomic fraction of mobile hydrogen, f_{Hm}, although f_{Hm} is slightly higher than $Y_{l,calculated}$.

The correspondence of the minimum of η with the maximum of f_{Hm} in the constant temperature experiments would not directly guarantee that the metaplast is fully quantified by mobile hydrogen. The situation would be also the same for the liquid fraction predicted by FG-DVC model. The Einstein coefficients, k_E, obtained empirically by fitting Eq. (1) to the observed viscosity data were 5.0 and 5.9 respectively for the model-predicted liquid fraction and the mobile hydrofgen. It would be possible to explain such rather high values assuming liquid entrapped with large agglomerates of solid particles. However, the values of k_E could be reduced with the assumption that both the mobile hydrogen and the liquid fraction are a portion of the metaplast. Also, determination of the coefficient, k_E, and the metaplast fraction, ϕ_l, depend on the viscosity of the metaplast, η_l. Since there have so far been reported a very limited number of data, further experiments are still needed to obtain η_l particularly in a range of temperature up to 750 K.

NOMENCLATURE

a	diameter of needle, [m]
b	diameter of pellet, [m]
E_V	activation energy for liquid viscosity, [$kJ \cdot mol^{-1}$]
f_{Hm}	atomic fraction of mobile hydrogen, [-]
g	gravitational acceleration, [$m \cdot s^{-2}$]
G	elastic modulus, [Pa]
h	net needle penetration depth, [m]
H	vertical position of needle tip, [m]
H_0	initial height of pellet, [m]
H_P	height of pellet at *t*, [m]
H_t	vertical distance from needle tip to pellet bottom, [m]
$I_G(0)$	initial signal intensity for Gaussian decay component, [-]
$I_L(0)$	initial signal intensity for exponential decay component, [-]
k_i	reaction rate constant, [s^{-1}]
k_E	Einstein coefficient in Eq.(1), [-]
m	mass of needle, [g]
Q	volumetric ratio of pellet at t to initial pellet, [-]
R	gas constant, [$kJ \cdot mol^{-1} \cdot K^{-1}$]
T	temperature, [K]
T_h	holding temperature, [K]
t	time in general or time after T_h is attained, [s]
Y_i	extract by solvent, [-]
γ	strain, [$rad \cdot s^{-1}$]
η	viscosity in general or apparent viscosity of coal pellet, [Pa·s]
η_l	apparent viscosity of liquid, [Pa·s]
η_{l0}	pre-exponential factor for liquid viscosity equation, [Pa·s]
σ	stress, [Pa]
τ	relaxation time defined by η/G, [s]
ϕ_c	critical volume fraction of solid in Mooney's equation, [-]
ϕ_{Po}	initial mass fraction of pyridine extract in semicoke/extract mixture, [-]
ϕ_P	mass fraction of pyridine extract in semicoke/extract mixture at a given temperature, [-]
ϕ_s	volume fraction of solid, [-]
ω	frequency, [s^{-1}]

REFERENCES

[1] Solomon, P. R. and Hamblen, D.G., *Chemistry of Coal Conversion*, Schlosberg, R. H., Ed., Plenum Pub., New York, Chap. 5, pp. 121-251, (1985)
[2] Helble, J. J. and Salofim, A. F., *Combust. Flame*, **76**, p. 183, (1989)

[3] Sero, M. A., Solomon, P. R., Bassilakis, R., and Suuberg, E. M., *Prepr. Am. Chem. Soc. Div. Fuel Chem.*, **34**(1), p. 9, (1989)
[4] Marsh, H. and Neavel, R. C., *Fuel*, **59**, p. 511, (1980)
[5] Lowry, H. H., *Chemistry of Coal Utilization*, Wiley, New York, (1963)
[6] ASTM D 2639-74, reapproved, (1980)
[7] Stevens, J. N., *CSIRO Division of Coal Research, Investigation Report*, No.12, (1957)
[8] Soth, G. C. and Russel, C. C., *Proc. Am. Soc. Test. Mater.*, **43**, p. 1176, (1943)
[9] Deng, C. R., Sanada, Y., and Chiba, T., *J. Fuel Soc. Jpn.*, **68**, p. 728, (1989)
[10] Lynch, L. J., Webster, R., Sakurovs, R., and Maher, T. P., *Fuel*, **67**, p. 579, (1988)
[11] Sakurovs, R., Lynch, L. J., Maher, T. P., and Banerjee, R. N., *Energy & Fuels*, **1**, p. 167, (1987)
[12] Eirich, F. R., Ed., *Rheology: Theory and Applications*, Academic Press, New York, (1958)
[13] Turi, E.A., Ed., *Thermal Characterization of Polymeric Materials*, Academic Press, London, (1981)
[14] Nomura, S., Kato, K., Komaki, I., Fujioka, Y., Saito, K., and Yamaoka, I., *Fuel*, **78**, p. 1583, (1999)
[15] Van Krevelen, D. W., Huntiens, F. J., and Dormans, H. N. M., *ibid*, **35**, p. 462, (1956)
[16] Chermin, H. A. G. and Van Krevelen, D. W., *ibid*, **36**, p. 85, (1957)
[17] Solomon, P. R., Best, P. E., Yu, Z. Z., and Charpenay, S., *Energy & Fuels*, **6**, p. 143, (1992)
[18] Mooney, M. J., *Colloid Sci.*, **6**, p. 162, (1951)
[19] Fong, W. S., Khalil, Y. E., Peters, W. A., and Howard, J. B., *Fuel*, **65**, p. 195, (1986)
[20] Morotomi, H., Suzuki, N., Miyazu, T., and Shimura, M., *J. Fuel Soc. Jpn.*, **53**, p. 779, (1974)
[21] Lewis, T. B. and Nielsen, L. E., *Trans. Soc. Rheol.*, **12**, p. 421, (1968)
[22] Einstein, A., *Ann. Physik*, **34**, p. 591, (1911)
[23] Soc. Polymer Sci., Jpn., *Rheology Measurements*, Kyoritsu Pub. Co., Tokyo, p. 129, (1965)
[24] Matsuoka, K., Hayashi, J. -i., and Chiba, T., *Iron & Steel Inst. J. Int.*, **37**, p. 566, (1997)
[25] Denma, D., *Master Thesis*, Graduate School of Eng., Hokkaido Univ., (1998)
[26] Hayashi, J. -i., Denma, D., Takahashi, H., Kumagai, H., and Chiba, T., *Fuel*, **79**, p. 391, (2000)
[27] Miyazawa, K., Yokono, T., and Sanada, Y., *Carbon*, **17**, p. 223, (1979)
[28] Maroto-Valer, M. M., Andresen, J. M., and Snape, C. E., *Energy & Fuels*, **11**, p. 236, (1997)
[29] Lynch, L. J. and Webster, D. S., *ACS Symp. Ser.*, **230**, p. 353, (1983)
[30] Lynch, L. J., Webster, D. S., and Barton, W. A., *Advances in Magnetic Resonance, Vol.12*, Waugh, J. S., Ed., Academic Press, New York, p. 385, (1988)

II. Coal Combustion

Chapter 4

ASH BEHAVIOR IN HIGH-TEMPERATURE PULVERIZED COAL COMBUSTION

*Ichiro Naruse**

1. INTRODUCTION

In order to enhance the net thermal efficiency in coal combustion/gasification systems, advanced coal combustion/gasification technologies at high temperature have been underdeveloped. Recently coal combustion/gasification technologies at high temperature have been applied not only to the Integrated Gasification Combined Cycle (IGCC) but also the Pulverized Coal Injection (PCI) technology in blast furnaces. Under the high temperature condition, however, a portion of ash components in coal is molten and/or evaporated. Hence, the ash molten or evaporated sometimes causes slagging and fouling problems. For coal gasifiers, a slag layer brings about plugging the furnace, and the ash deposit on the heat exchanger tubes makes the overall heat transfer coefficient reduce. For blast furnaces, on the other hand, the ash particulates may cause an increase of the pressure drop due to interaction of the molten ash with a coke lump near the raceway [1]. However, it is necessary for sophisticated development to elucidate such ash characteristics as melting, evaporating, condensing and depositing since those characteristics closely relate to performance.

Generally, ash behaviors are known to be influenced by coal types, reaction temperature and atmosphere, particle temperature, surface temperature of a heat exchanger, material of its surface, flow dynamics, heat transfer and so forth. Benson *et al.* [2] summarized phenomena of the ash depositions in coal combustion, and voluminous works were referred to in their summary. For instance, advanced analytical methods of the ash compositions in coal particles with a CCSEM (Computer-Controlled Scanning Electron Microscope) [3, 4] system, formation of the ash intermediates which consist of

* Department of Ecological Engineering, Toyohashi University of Technology, 1-1 Hibarigaoka, Tempaku-cho, Toyohashi 441-8580. Japan (naruse@eco.tut.ac.jp).

gases, liquids and solids [5], the deposit initiation [6, 7], the deposition characteristics and deposit growth [8], properties of the deposit [9, 10], model of the ash behaviors [11] and so forth are referred to there [2]. Even from those references, however, the precise and quantitative knowledge in high-temperature coal combustion or gasification has not been obtained yet due to the complexity in determining the relationship between the coal reaction and the ash behaviors quantitatively. Especially for the slagging problems, studies on the melting characteristics of coal ash [12, 13, 14] and the influence of additives like CaO and FeO on lowering of the ash melting point [15, 16, 17] have also been reported.

In this chapter, fundamental reaction characteristics of a single coal particle, which is trapped on/in a high-temperature molten slag, are mentioned first. This is because the phenomenon may initiate ash deposition on the furnace wall. Second, fundamentals of ash deposition behaviors under high-temperature reaction conditions of coal are introduced. Additionally, the ash deposition characteristics in the practical systems are also mentioned, based on some literature. Finally, emissions of trace metal compounds in combustion processes and its control technologies are introduced briefly. This is one of recent issues on the ash worldwide.

2. Fundamental Reaction Characteristics of Coal Particles Trapped on/in High Temperature Molten Slag [18]

Under high-temperature reaction conditions, coal rapidly evolves a volatile matter, and fixed carbon also burns. At the same time, some ash components in coal may melt inside a particle. This phenomenon suggests that both the reaction behavior and phase condition of a single coal particle may affect the ash deposition characteristics, especially for slagging behavior. From this viewpoint, fundamental behaviors of combustion/gasification of a single coal particle in a molten slag layer are introduced here.

The fundamental behaviors of combustion/gasification of a coal particle trapped on the molten slag layer were studied by using a laboratory scale batch reactor. Figure 1 illustrates a schematic diagram of an experimental apparatus. It basically consists of a high temperature furnace made of alumina, which is heated by propane/oxygen enriched air premixed flame, and a reaction tube made of alumina. The reaction tube is constructed as double in which the reactant flows down through the inner tube and the reaction product flows up through the annular part. Air and CO_2 were used as the reactants to study the reaction phenomena under the combustion and gasification conditions, respectively. Gas sampled is guided to several analyzers through a tube at the entrance of which the product gas is diluted and quenched by nitrogen. Over the inner tube an observation window is provided to allow optical measurement of the surface temperature. An injection port of single coal particle is also mounted. The single coal particle is introduced into the reactor by nitrogen pulse. The signal obtained by the analyzer is